Setup of a Graphical User Interface Desktop for Linux Virtual Machine on Cloud Platforms

By

Dr. Hidaia Mahmood Alassouli

Hidaia_alassouli@hotmail.com

Setup of a Graphical User Interface Desktop for Linux Virtual Machine on Cloud Platforms

Written by Dr. Hidaia Mahmood Alassouli.

1. Introduction:

Cloud Platforms provide VM images in the Linux OS as well. Linux has always been operated via terminal or shell through a keyboard and a terminal. Even with GUIs around, Linux continues to be operated from the shell.

Linux VMs are also operated from the command line of your desktop via an SSH (secure shell) connection. They do not have a desktop environment or GUI installed by default. For Windows users migrating to Linux, a desktop environment would be more convenient to operate. Hence, various desktop environments can be set up on a Linux VM.

Mostly we need to have Graphical User Interface GUI on the Linux Virtual Machine instance and to use Internet browser on it.

This report will talk about the steps to install minimum required User Interface on VM (virtual machine) with Web Browser. We will work on installing a desktop environment on a Linux Virtual Machine on different Cloud Platforms. The book consists from the following sections:

1. Generating SSH key for auto log in to Linux server.
2. Creating Google Cloud Linux Virtual Machine.
3. Logon to the Linux Virtual Machine
4. Installing VNC server.
5. Installing XRDP server.
6. Installing a Graphical User Interface (GUI) for Linux Google Cloud instance and connecting to the server through VNC or RDP connection.
7. Quick guide to create a Linux virtual machine in Cloudsigma.
8. Quick guide to create a Linux Virtual Machine in the Microsoft Azure portal.
9. Quick guide to create a Linux Virtual Machine in Amazon AWS.

2. Generating SSH key for auto log in to Linux server:

1. Start by downloading and installing "PuTTY installer".

2. Select PuttyGen and press "Enter". Once "PuttyGen" is opened press the "Generate" button. Then you have to move your mouse in the empty area randomly until your key is generated. After this append your name or email to the Key comment. This step is optional but I highly recommend it. When you are working in teams it's easier for a system administrator to manage access based on SSH Keys. Protect your SSH key with a password by filling in the "Key passphrase" field.

3. Let's generate SSH key using Putty Key Generator PuttyGen tool. This is the screen of PuttyGen tool.

4. Click generate. Move your mouser in the empty space to generate the key.

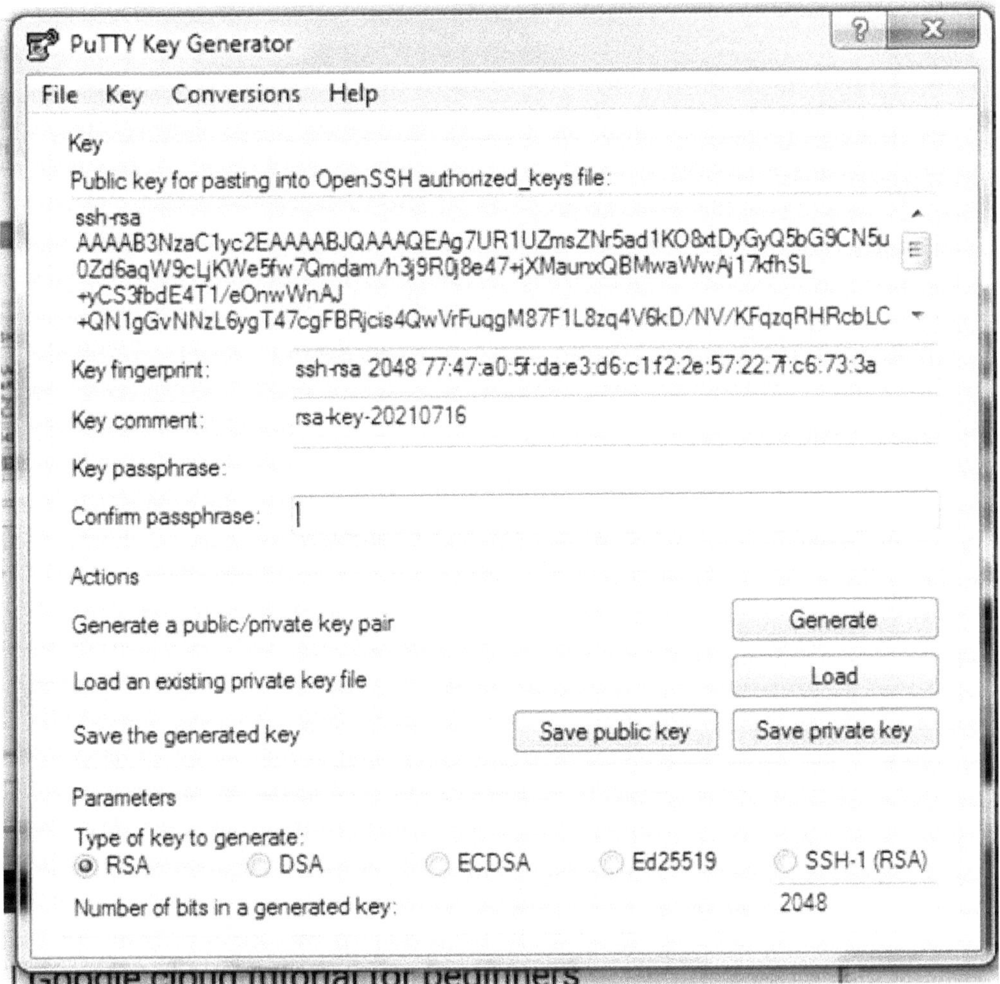

5. Give name to key comment. I chose my name "hidaia".

6. Now we have to save it in different formats. Firstly, we'll start by pressing the "Save public key" button; navigate to your convenient folder and save it as "hidaia-public-key.pub". Secondly, we'll press the "Save private key" button and save it as "Hidaia-private-key.ppk". And lastly, since some software require it in OpenSSH format, click on the "Conversions" menu located at the top of the window and select "Export OpenSSH key", then save it as "Hidaia-openssh-key.openssh" and we are done with this part. Let's close PuttyGen for now.

PuTTY Key Generator
File Key Conversions Help
Key
Public key for pasting into OpenSSH authorized_keys file:
ssh-rsa
AAAAB3NzaC1yc2EAAAABJQAAAQEAg7UR1UZmsZNr5ad1KO8xtDyGyQ5bG9CN5u
0Zd6aqW9cLjKWe5fw7Qmdam/h3j9R0j8e47+jXMaunxQBMwaWwAj17kfhSL
+yCS3fbdE4T1/eOnwWnAJ
+QN1gGvNNzL6ygT47cgFBRjcis4QwVrFuqgM87F1L8zq4V6kD/NV/KFqzqRHRcbLC
Key fingerprint: ssh-rsa 2048 77:47:a0:5f:da:e3:d6:c1:f2:2e:57:22:7f:c6:73:3a
Key comment: hidaia
Key passphrase:
Confirm passphrase:
Actions
Generate a public/private key pair
Generate
Load an existing private key file
Load
Save the generated key
Save public key
Save private key
Parameters
Type of key to generate:
RSA
DSA
ECDSA
Ed25519
SSH-1 (RSA)
Number of bits in a generated key: 2048

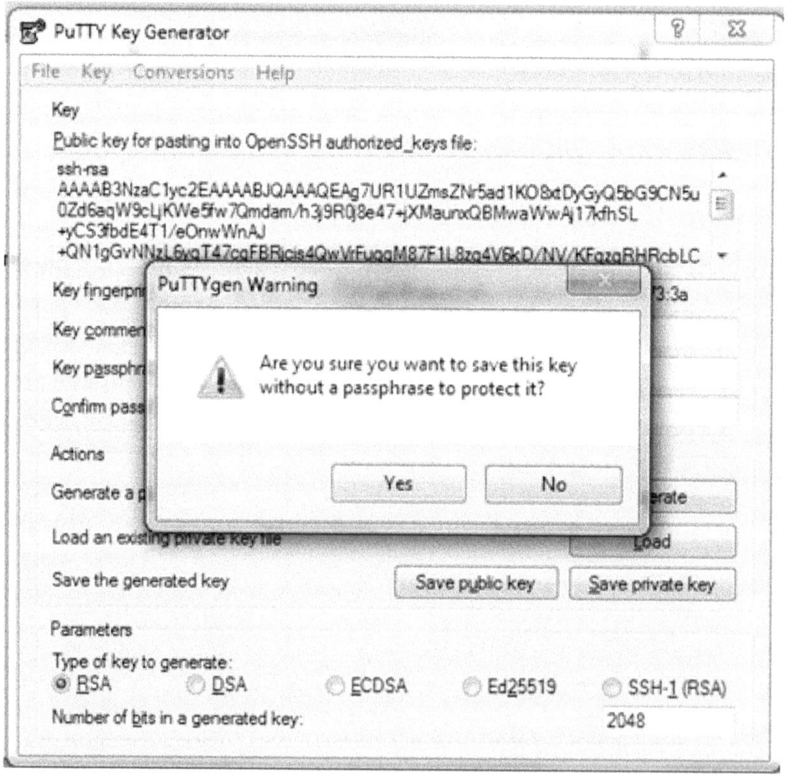
PuTTY Key Generator
File Key Conversions Help
Key
Public key for pasting into OpenSSH authorized_keys file:
ssh-rsa
AAAAB3NzaC1yc2EAAAABJQAAAQEAg7UR1UZmsZNr5ad1KO8xtDyGyQ5bG9CN5u
0Zd6aqW9cLjKWe5fw7Qmdam/h3j9R0j8e47+jXMaunxQBMwaWwAj17kfhSL
+yCS3fbdE4T1/eOnwWnAJ
PuTTYgen Warning
Are you sure you want to save this key without a passphrase to protect it?
Yes
No
Save the generated key
Save public key
Save private key
Parameters
Type of key to generate:
RSA
DSA
ECDSA
Ed25519
SSH-1 (RSA)
Number of bits in a generated key: 2048

7. There are many ways to auto log in with your ssh key using Putty. One way is to manage the requests for your SSH keys with Pageant. Run "Putty Pageant" tool. Select "Add key" and navigate to the path where you have saved your SSH Key and double click it. Input your passphrase. Then close the Pageant window. Putty will use the key information in Pageant when creating SSH connection to the server.

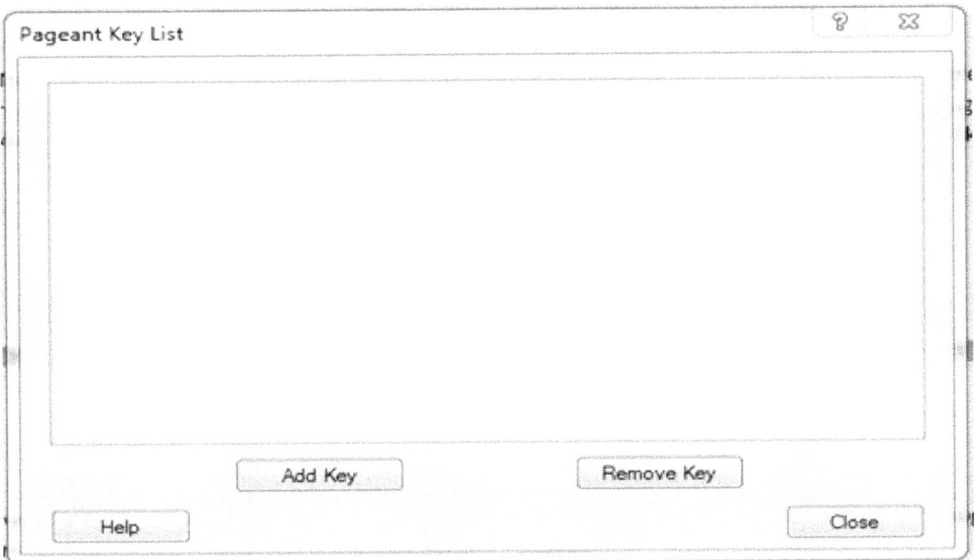

8. Setup the virtual Linux server with the SSH keys.

9. Using Putty tool, create SSH connection to public address of the Linux server. Putty tool uses the key credentials saved Pageant when creating the connection.

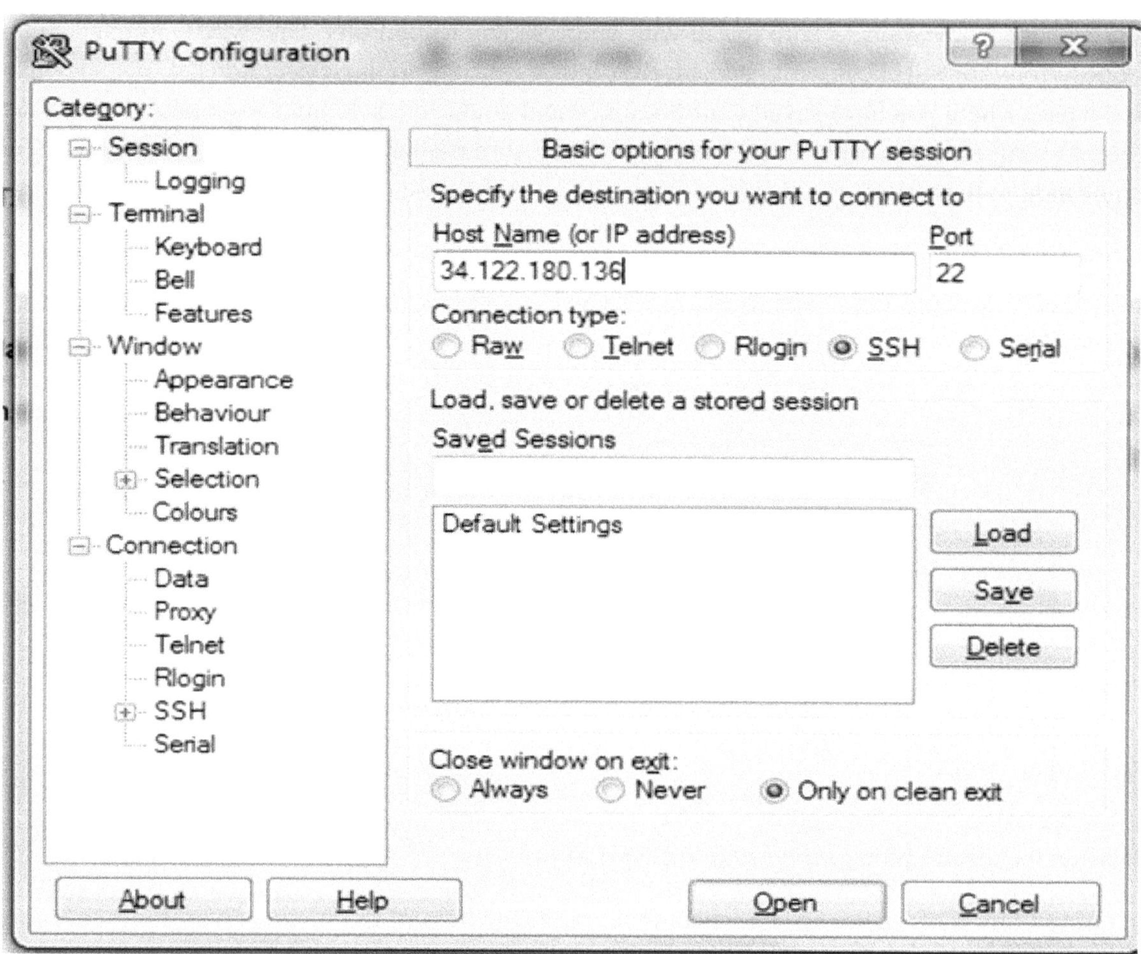
PuTTY Configuration
Category:
Session
Logging
Terminal
Keyboard
Bell
Features
Window
Appearance
Behaviour
Translation
Selection
Colours
Connection
Data
Proxy
Telnet
Rlogin
SSH
Serial
Basic options for your PuTTY session
Specify the destination you want to connect to
Host Name (or IP address)
Port
34.122.180.136
22
Connection type:
Raw
Telnet
Rlogin
SSH
Serial
Load, save or delete a stored session
Saved Sessions
Default Settings
Load
Save
Delete
Close window on exit:
Always
Never
Only on clean exit
About
Help
Open
Cancel

3. Creating Google Cloud Linux Virtual Machine:

I will show here the steps to create Linux Virtual Server using Google cloud as the setup of Linux virtual server quite complicated

1. Go to console.cloud.google.com. Login using your login details.

2. Create compute engine virtual server instance.

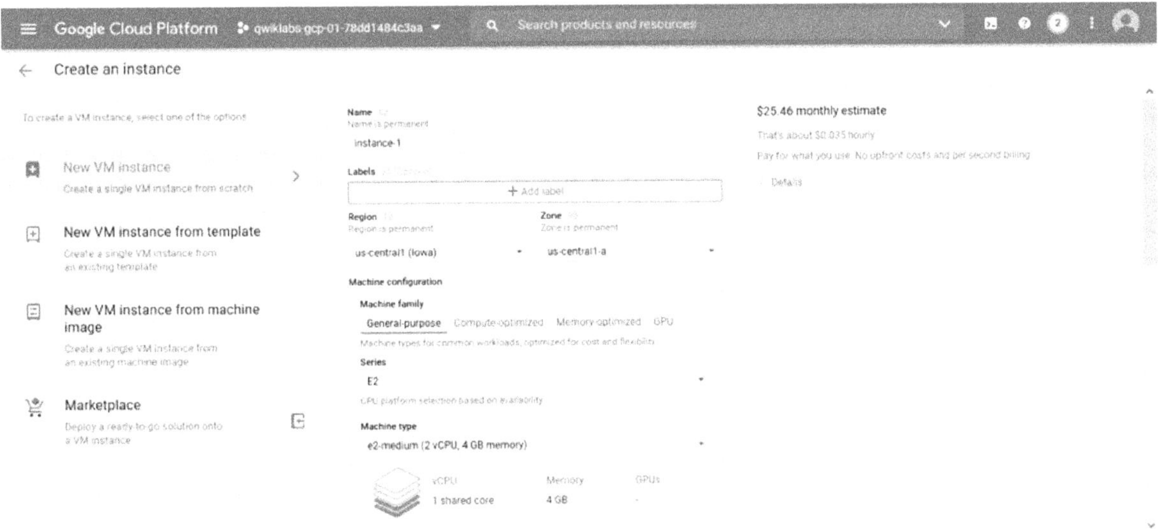

3. Choose to create Linux Virtual Server. I chose Ubuntu server.

4. Expand "Networking, Security, Networking and sole Tenancy" section. And go to security part to define the SSH public key.

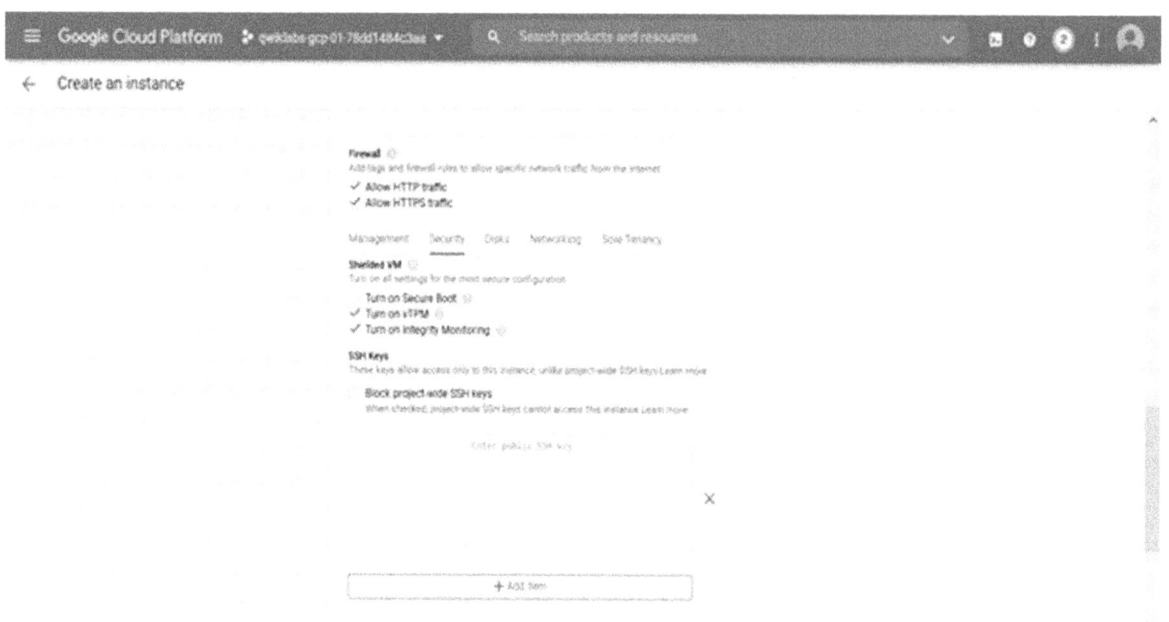

5. Let's generate SSH key using Putty Key Generator PuttyGen tool. This is the screen of PuttyGen tool.

6. Click generate. Move your mouse in the empty space to generate the key.

7. Give name to key comment. I chose my name “hidaia”. Then select “save private key”. You can setup passphrase or just leave it empty.

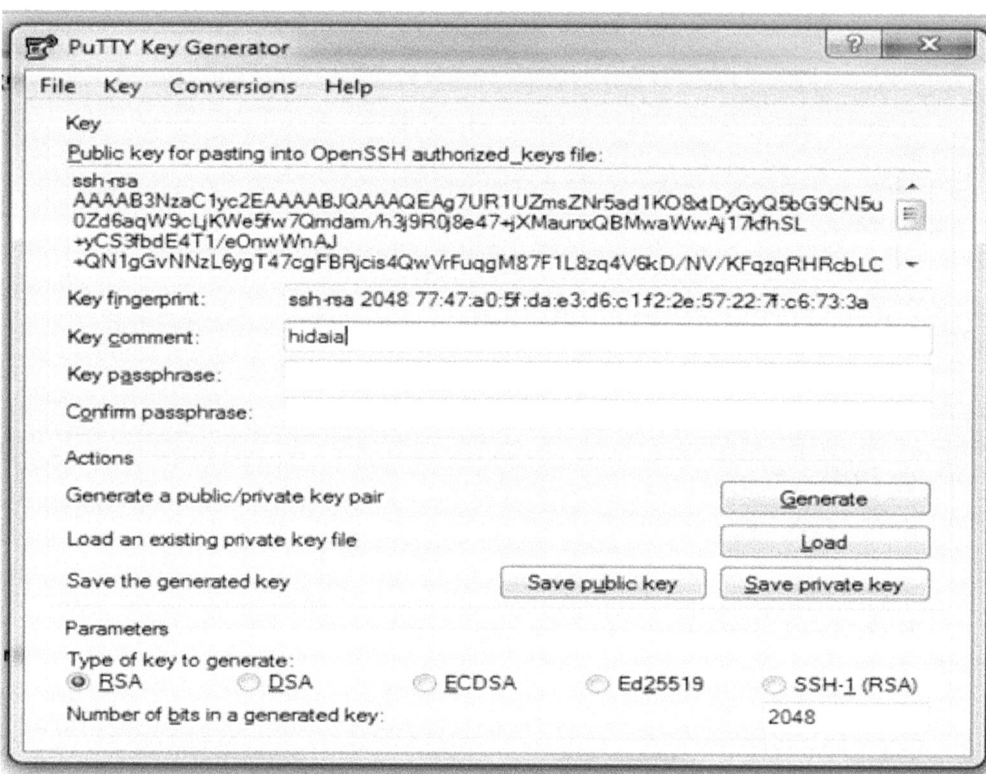

8. Save the generated public and private key file.

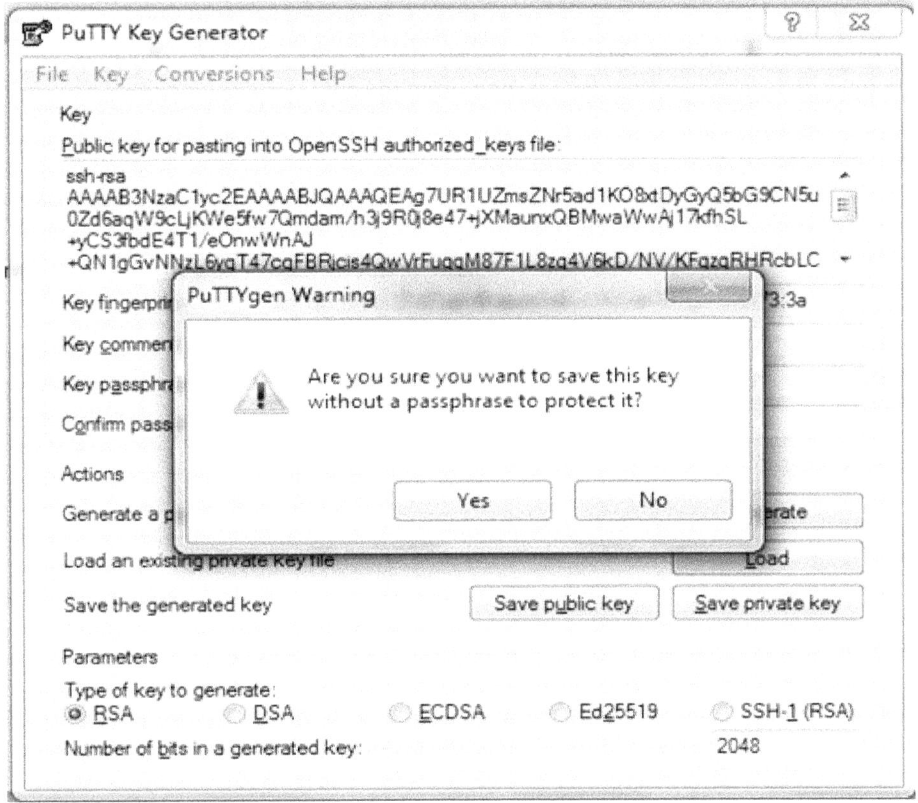

9. Copy the key and paste it in Google cloud virtual server security SSH Keys section.

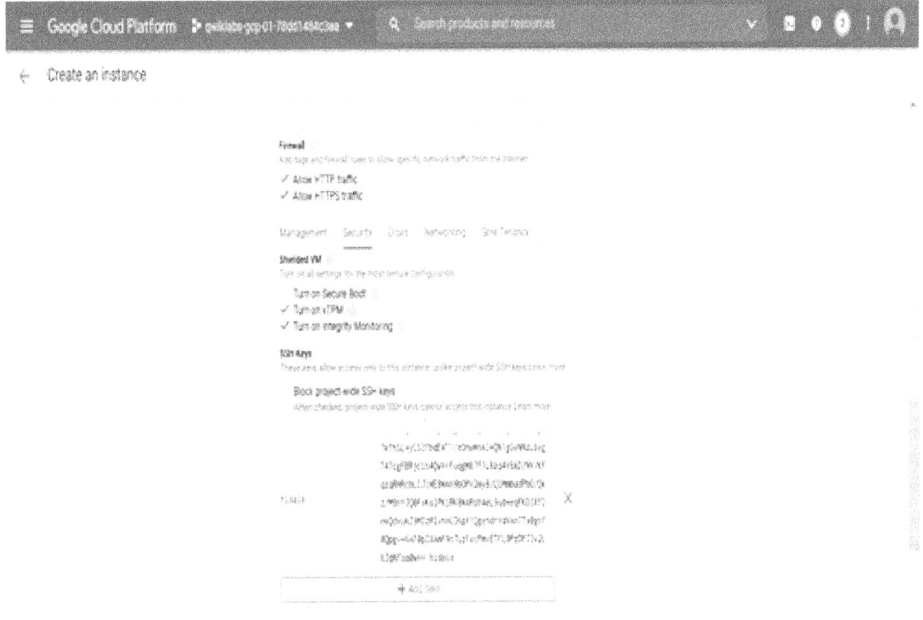

10. Click on Create button. The Linux virtual server instance will be generated. Here screenshot of the created instance. It will be shown the public IP of the server you are going to log onto it. But we don't have the login credentials as the username and password. The login will be through the SSH key.

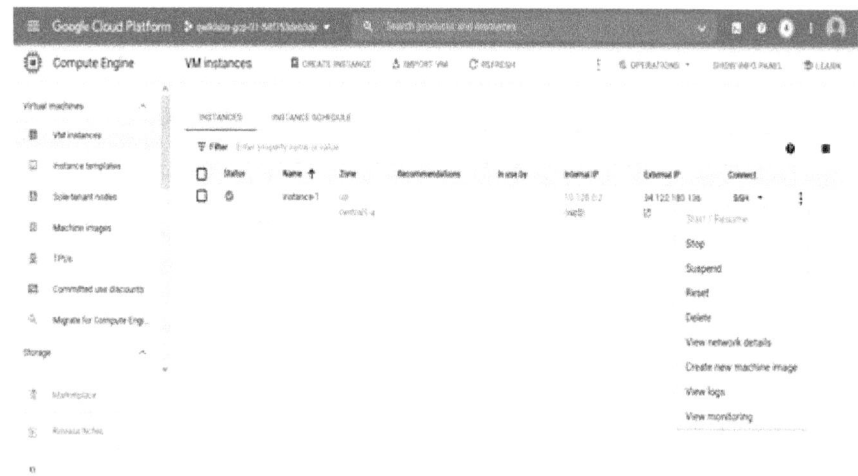

11. If you did not provide the machine SSH key during setup, you can edit the machine SSH key from the virtual machine properties menu and enter your generated SSH key through PuTTY.

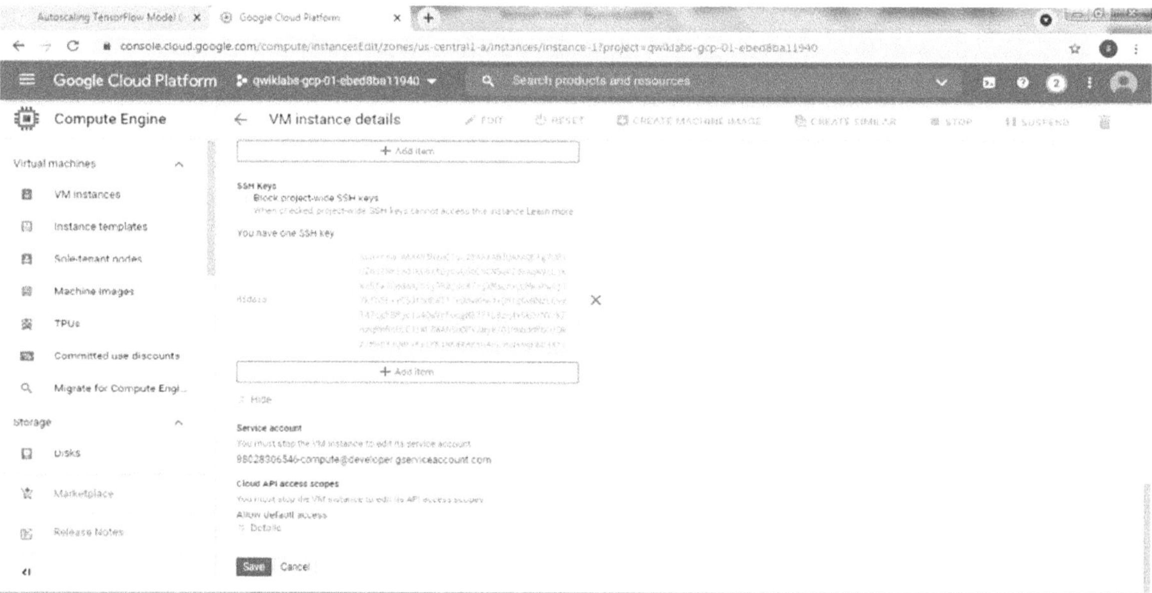

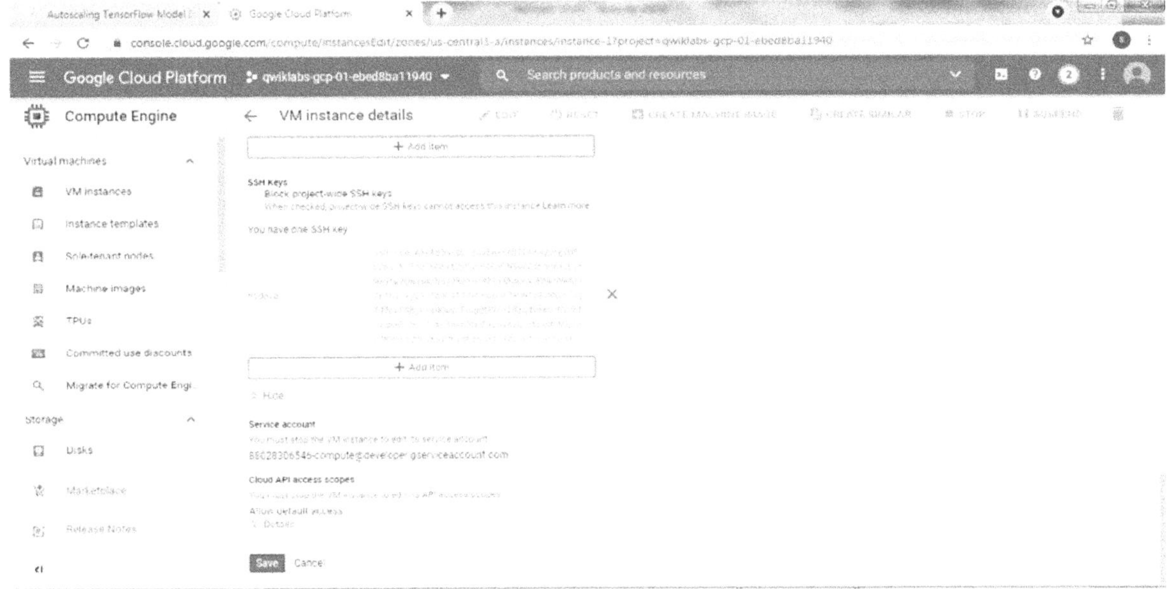

12. Otherwise, if you did not provide at all machine SSH key, the Google Cloud Platform will generate default SSH key and you can login through it using Google Cloud browser window.

4. Logon to the Linux Virtual Machine:

There are many ways to access the created Linux Virtual Machine.

a) Logon to Linux Virtual Machines through Google Cloud browser window:

1. This is the screenshot of the instance I created for the Linux Virtual Server

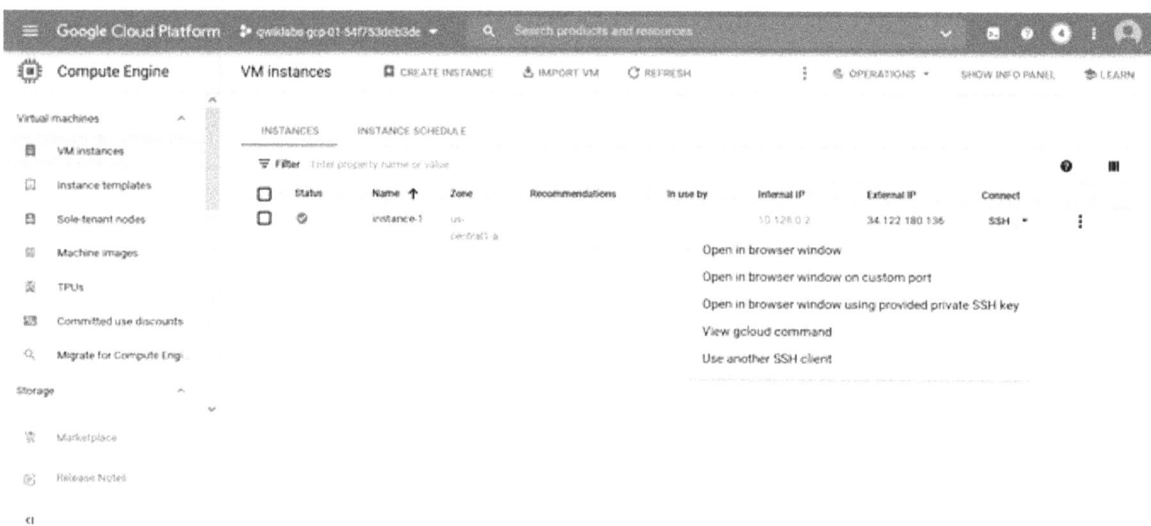

2. Choose the first option which is to login to the server through opening the SSH connection in browser window. You will get such as this pop-up window

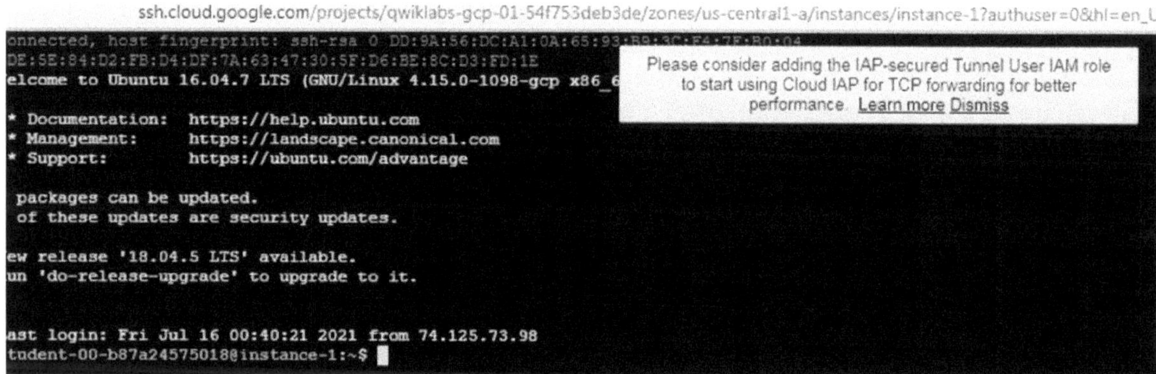

3. You are logged automatically using the SSH key defined during machine setup or the SSH key generated by Google Cloud if you did not define SSH key.

4. Here some commands I tested with their output:

$ ls

$ free

```
student-00-b87a24575018@instance-1:~$ ls
student-00-b87a24575018@instance-1:~$ dir
student-00-b87a24575018@instance-1:~$ free
              total        used        free      shared  buff/cache   available
Mem:        4038480      108924     3592796        5668      336760     3671412
Swap:             0           0           0
student-00-b87a24575018@instance-1:~$
```

- sudo apt-get update command to update the system packages

  $ sudo apt-get update

```
student-00-b87a24575018@instance-1:~$ sudo apt-get update
Hit:1 http://us-central1.gce.archive.ubuntu.com/ubuntu xenial
Get:2 http://us-central1.gce.archive.ubuntu.com/ubuntu xenial-
Get:3 http://us-central1.gce.archive.ubuntu.com/ubuntu xenial-
Get:4 http://security.ubuntu.com/ubuntu xenial-security InRele
Get:5 http://us-central1.gce.archive.ubuntu.com/ubuntu xenial/universe amd64 Packages [7,532 kB]
Get:6 http://archive.canonical.com/ubuntu xenial InRelease [11.5 kB]
Get:7 http://us-central1.gce.archive.ubuntu.com/ubuntu xenial/universe Translation-en [4,354 kB]
Get:8 http://us-central1.gce.archive.ubuntu.com/ubuntu xenial/multiverse amd64 Packages [144 kB]
Get:9 http://us-central1.gce.archive.ubuntu.com/ubuntu xenial/multiverse Translation-en [106 kB]
Get:10 http://us-central1.gce.archive.ubuntu.com/ubuntu xenial-updates/main amd64 Packages [2,049 kB]
Get:11 http://us-central1.gce.archive.ubuntu.com/ubuntu xenial-updates/main Translation-en [482 kB]
Get:12 http://us-central1.gce.archive.ubuntu.com/ubuntu xenial-updates/universe amd64 Packages [1,219 kB]
Get:13 http://us-central1.gce.archive.ubuntu.com/ubuntu xenial-updates/universe Translation-en [358 kB]
Get:14 http://us-central1.gce.archive.ubuntu.com/ubuntu xenial-updates/multiverse amd64 Packages [22.6 kB]
Get:15 http://us-central1.gce.archive.ubuntu.com/ubuntu xenial-updates/multiverse Translation-en [8,476 B]
Get:16 http://us-central1.gce.archive.ubuntu.com/ubuntu xenial-backports/main amd64 Packages [9,812 B]
Get:17 http://us-central1.gce.archive.ubuntu.com/ubuntu xenial-backports/main Translation-en [4,456 B]
Get:18 http://us-central1.gce.archive.ubuntu.com/ubuntu xenial-backports/universe amd64 Packages [11.3 kB]
Get:19 http://us-central1.gce.archive.ubuntu.com/ubuntu xenial-backports/universe Translation-en [4,476 B]
Get:20 http://security.ubuntu.com/ubuntu xenial-security/main amd64 Packages [1,648 kB]
Get:21 http://security.ubuntu.com/ubuntu xenial-security/main Translation-en [380 kB]
Get:22 http://archive.canonical.com/ubuntu xenial/partner amd64 Packages [2,696 B]
Get:23 http://security.ubuntu.com/ubuntu xenial-security/universe amd64 Packages [785 kB]
Get:24 http://security.ubuntu.com/ubuntu xenial-security/universe Translation-en [225 kB]
Get:25 http://security.ubuntu.com/ubuntu xenial-security/multiverse amd64 Packages [7,864 B]
Get:26 http://security.ubuntu.com/ubuntu xenial-security/multiverse Translation-en [2,672 B]
Get:27 http://archive.canonical.com/ubuntu xenial/partner Translation-en [1,556 B]
Fetched 19.7 MB in 3s (5,196 kB/s)
Reading package lists... Done
student-00-b87a24575018@instance-1:~$
student-00-b87a24575018@instance-1:~$
```

Please consider adding the IAP-secured Tunnel User IAM
to start using Cloud IAP for TCP forwarding for better
performance. Learn more Dismiss

- sudo apt-get install command will install any package you want

  $ sudo apt-get install htop

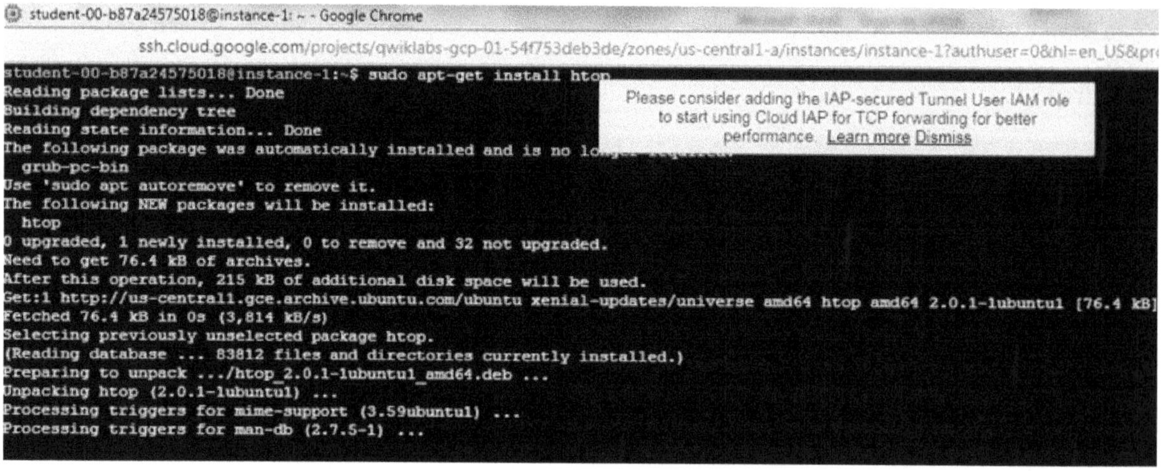

student-00-b87a24575018@instance-1: ~ - Google Chrome

ssh.cloud.google.com/projects/qwiklabs-gcp-01-54f753deb3de/zones/us-central1-a/instances/instance-1?authuser=0&hl=en_US&pr

```
student-00-b87a24575018@instance-1:~$ sudo apt-get install htop
Reading package lists... Done
Building dependency tree
Reading state information... Done
The following package was automatically installed and is no lo
  grub-pc-bin
Use 'sudo apt autoremove' to remove it.
The following NEW packages will be installed:
  htop
0 upgraded, 1 newly installed, 0 to remove and 32 not upgraded.
Need to get 76.4 kB of archives.
After this operation, 215 kB of additional disk space will be used.
Get:1 http://us-central1.gce.archive.ubuntu.com/ubuntu xenial-updates/universe amd64 htop amd64 2.0.1-1ubuntu1 [76.4 kB]
Fetched 76.4 kB in 0s (3,814 kB/s)
Selecting previously unselected package htop.
(Reading database ... 83812 files and directories currently installed.)
Preparing to unpack .../htop_2.0.1-1ubuntu1_amd64.deb ...
Unpacking htop (2.0.1-1ubuntu1) ...
Processing triggers for mime-support (3.59ubuntu1) ...
Processing triggers for man-db (2.7.5-1) ...
```

Please consider adding the IAP-secured Tunnel User IAM role to start using Cloud IAP for TCP forwarding for better performance. Learn more Dismiss

5. Instead of using SSH key every time to login, you can create username and password credentials which will be easier to be used as login details using any SSH client. Fellow these steps

- Change your login to be as root super user using command sudo su

  $ sudo su

- Check what is your current username using command whami

  # whoami

- Edit the /etc/ssh/sshd_config ssh configuration file to allow authentication using username and password credentials,

  # vim /etc/ssh/sshd_config

  Note: To open a file using Vim, launch your terminal and type vim followed by the name of the file you want to edit or create. To save the file without exiting the editor, switch back to normal mode by pressing Esc, type :w and hit Enter. To save the file and exit the editor simultaneously, press Esc to switch to normal mode, type :wq and hit Enter. To exit the editor, without saving the changes, switch to normal mode by pressing Esc, type :q! and hit Enter.

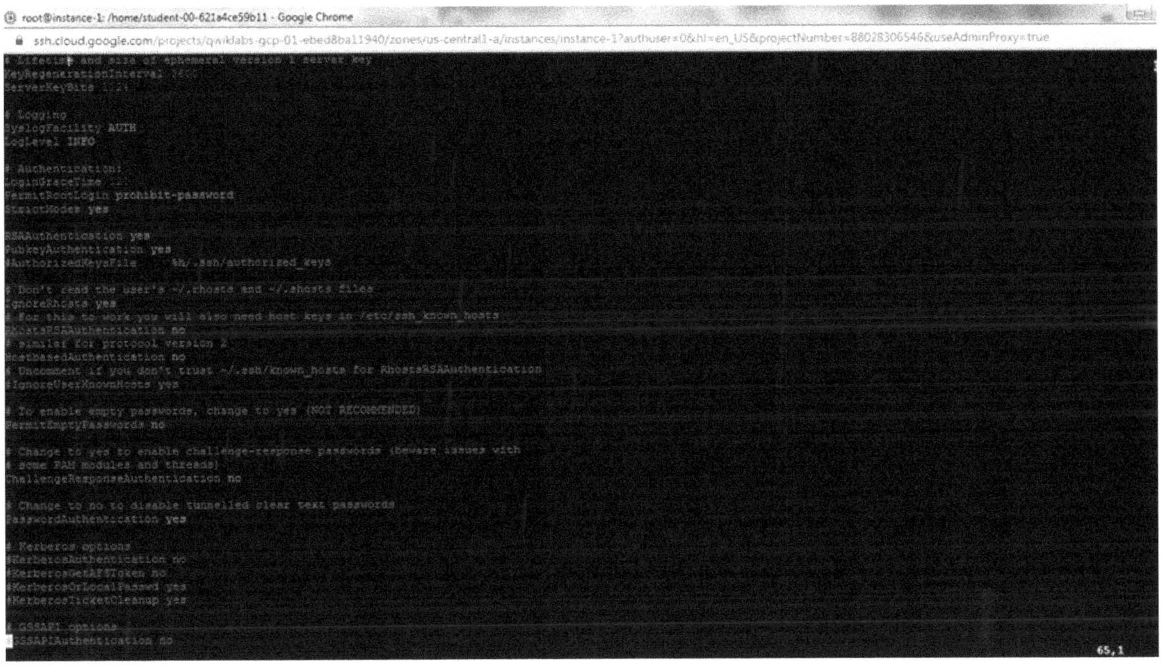

root@instance-1: /home/student-00-621a4ce59b11 - Google Chrome

```
# Lifetime and size of ephemeral version 1 server key
KeyRegenerationInterval 3600
ServerKeyBits 1024

# Logging
SyslogFacility AUTH
LogLevel INFO

# Authentication:
LoginGraceTime 120
PermitRootLogin prohibit-password
StrictModes yes

RSAAuthentication yes
PubkeyAuthentication yes
#AuthorizedKeysFile     %h/.ssh/authorized_keys

# Don't read the user's ~/.rhosts and ~/.shosts files
IgnoreRhosts yes
# For this to work you will also need host keys in /etc/ssh_known_hosts
RhostsRSAAuthentication no
# similar for protocol version 2
HostbasedAuthentication no
# Uncomment if you don't trust ~/.ssh/known_hosts for RhostsRSAAuthentication
#IgnoreUserKnownHosts yes

# To enable empty passwords, change to yes (NOT RECOMMENDED)
PermitEmptyPasswords no

# Change to yes to enable challenge-response passwords (beware issues with
# some PAM modules and threads)
ChallengeResponseAuthentication no

# Change to no to disable tunnelled clear text passwords
PasswordAuthentication yes

# Kerberos options
#KerberosAuthentication no
#KerberosGetAFSToken no
#KerberosOrLocalPasswd yes
#KerberosTicketCleanup yes

# GSSAPI options
#GSSAPIAuthentication no
```

- Restart the service sshd

  # service sshd restart

- Create username, as example halassouli using add user command. Specify the password and user information's.

  # adduser halassouli

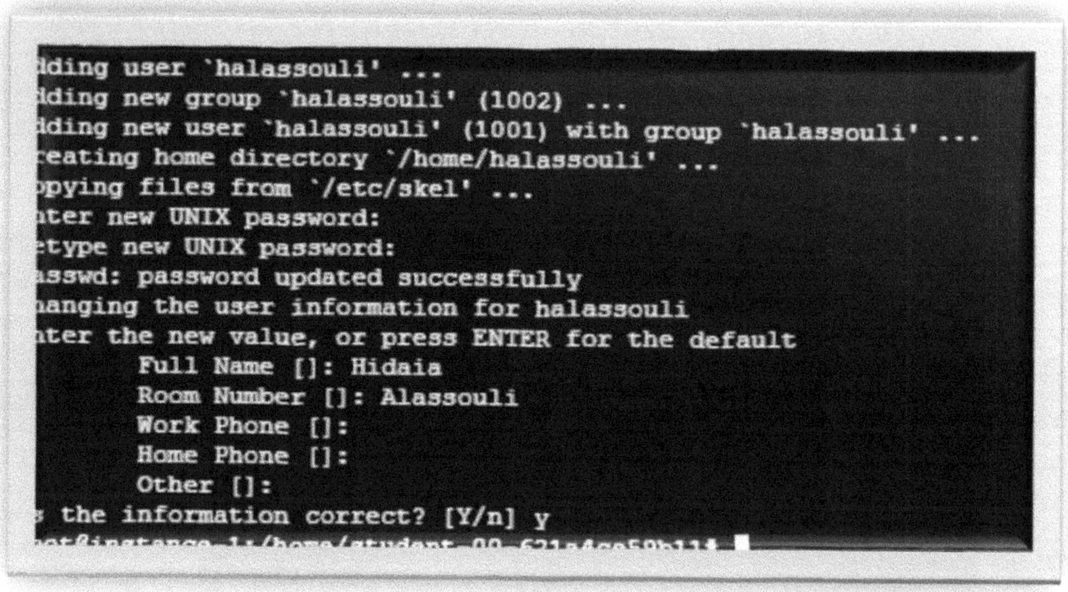

```
lding user `halassouli' ...
lding new group `halassouli' (1002) ...
lding new user `halassouli' (1001) with group `halassouli' ...
reating home directory `/home/halassouli' ...
opying files from `/etc/skel' ...
nter new UNIX password:
etype new UNIX password:
asswd: password updated successfully
hanging the user information for halassouli
nter the new value, or press ENTER for the default
        Full Name []: Hidaia
        Room Number []: Alassouli
        Work Phone []:
        Home Phone []:
        Other []:
s the information correct? [Y/n] y
```

- Let's try to login using PuTTY SSH client with the created username and password with the public IP of the server. Don't specify any SSH keys in this case.

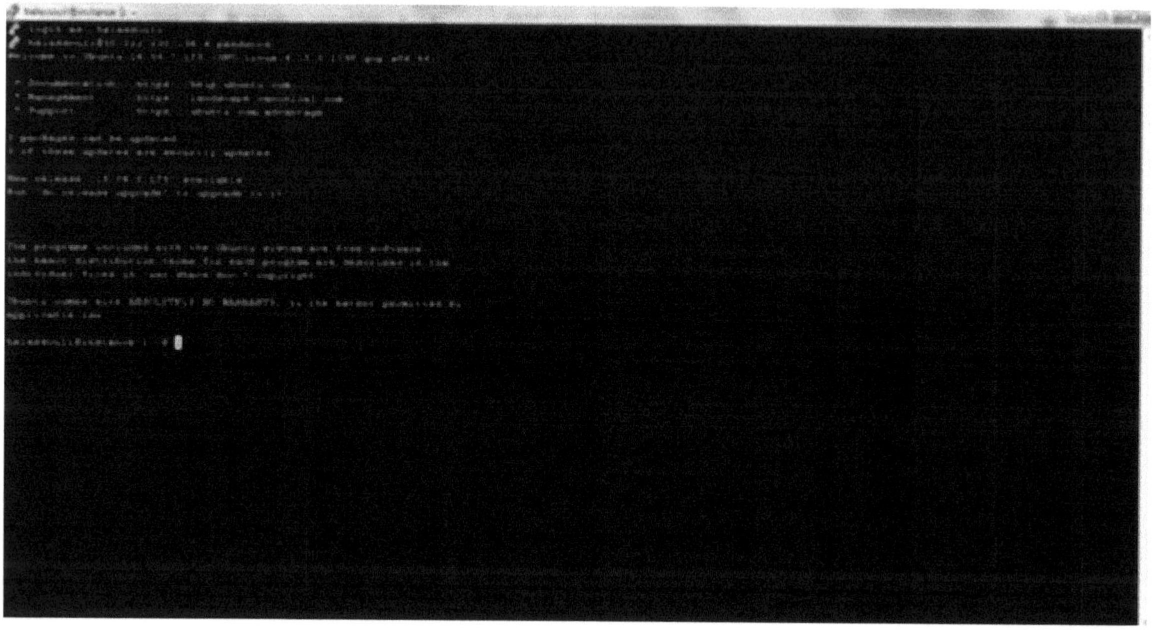

b) Logon to Linux Virtual Machines through PuTTY SSH connection using the SSH key:

In this case you must have the SSH key to login to the machine.

1. You must use Putty Pageant tool.

2. Select Add to add Key file saved for key when generated. Then close the Pageant window. PuTTY will use the key information in Pageant tool when creating SSH connection to the server.

3. Using Putty tool, create SSH connection to public address of the Linux server. Putty tool uses the key credentials saved Pageant when creating the connection.

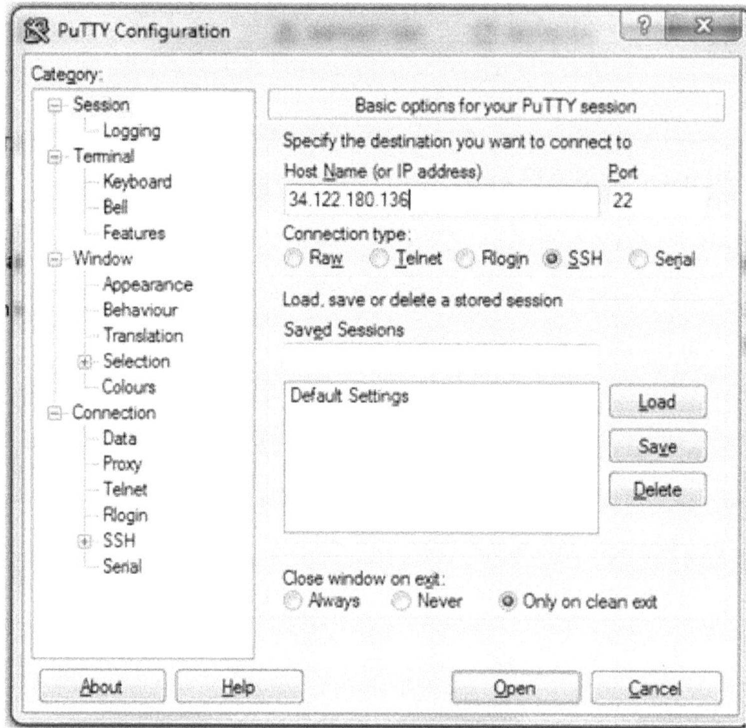

4. Instead of using Pageant tool to identify the SSH key, you can define the SSH private key in PuTTY tool directly from the Connection/SSH/Auth section.

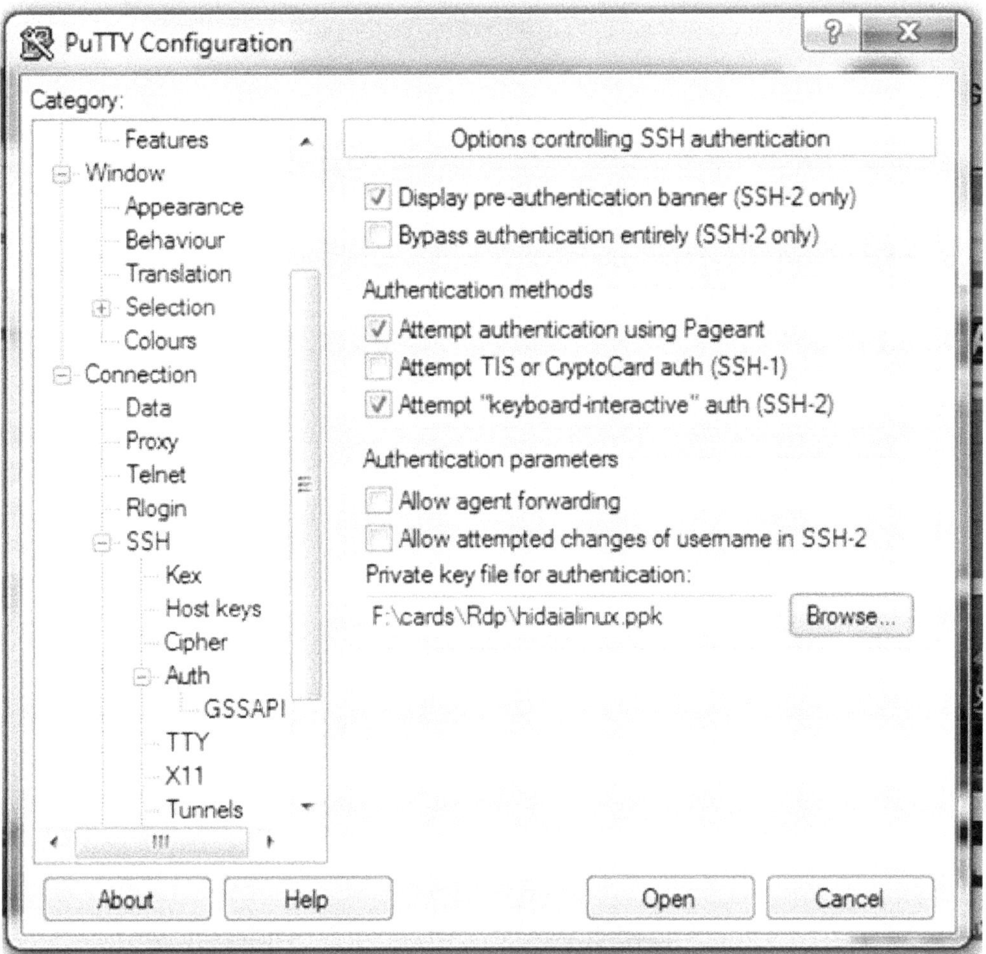

5. Provide the username for login which is the "key comment" you defined during the generation of the SSH key using the PuTTY key generator.

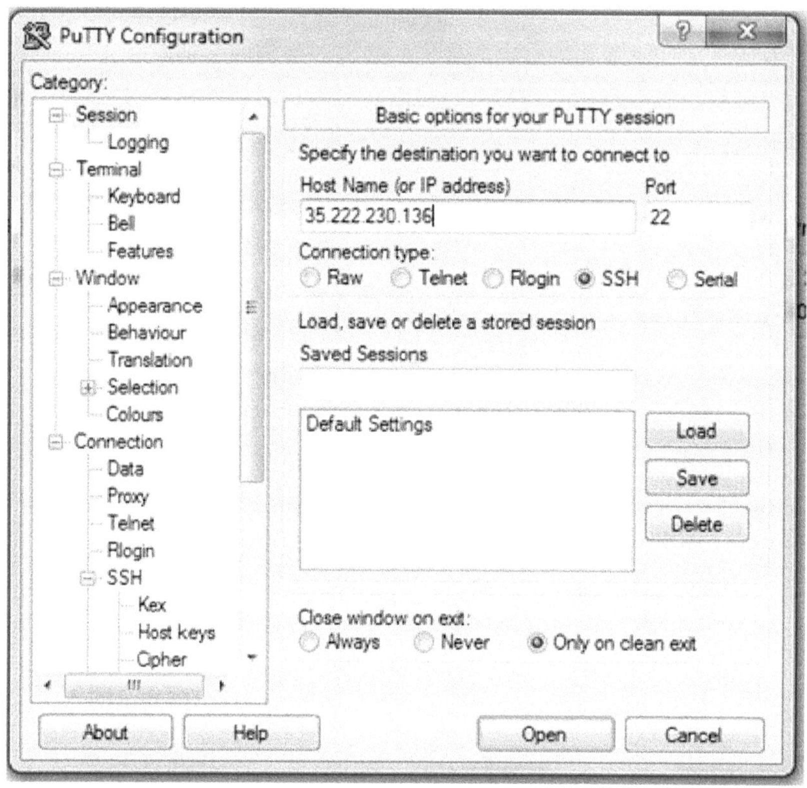

6. You can repeat the steps in previous section to create new username and password to be used as login credentials instead of SSH keys.

- Create username, as example hasooly using add user command. Specify the password and user information.

```
$ sudo adduser hasooly
```

```
student-00-a3ac361471ed@instance-1:~$ sudo adduser hasooly
Adding user `hasooly' ...
Adding new group `hasooly' (1003) ...
Adding new user `hasooly' (1002) with group `hasooly' ...
Creating home directory `/home/hasooly' ...
Copying files from `/etc/skel' ...
Enter new UNIX password:
Retype new UNIX password:
passwd: password updated successfully
Changing the user information for hasooly
Enter the new value, or press ENTER for the default
        Full Name []: hidaia
        Room Number []:
        Work Phone []:
        Home Phone []:
        Other []:
Is the information correct? [Y/n] y
student-00-a3ac361471ed@instance-1:~$
```

- To update also the root user, connect to the server using super root account. Then change the password of the root user by passwd command

  $ sudo su

  # passwd

  Give the new password of the root

```
root@instance-1:/home/student-00-a3ac361471ed# passwd
Enter new UNIX password:
Retype new UNIX password:
passwd: password updated successfully
root@instance-1:/home/student-00-a3ac361471ed#
```

5. Installing VNC server:

1. Login to Google Cloud console. Navigate the left menu to list the VM instances running on your project: Compute > Compute Engine > VM instances. Click on the Create Instance button to access the instance creation form. Repeat the steps in section a to create Linux virtual machine

2. Connect to the Linux virtual machine using the Google Cloud browser. You will be connected to the machine using the SSH key you defined during machine setup, or the default SSH keys provided by Google Cloud.

3. Update the Linux machine packages using the command "sudo apt-get update" command.

```
$ sudo apt-get  update

$ sudo apt-get  upgrade
```

4. Install tightvncserver

```
$ sudo apt-get install tightvncserver
$ sudo ufw allow 5901:5910/tcp
```

5. Now that our instance has a desktop environment let's make it accessible via VNC. Start the vncserver, and follow the directions to create a password

```
$ vncserver
```

```
student-00-621a4ce59b11@instance-1:~$ vncserver

You will require a password to access your desktops.

Password:
Verify:
Would you like to enter a view-only password (y/n)? y
Password:
Verify:
xauth:   file /home/student-00-621a4ce59b11/.Xauthority does not exist

New 'X' desktop is instance-1:1

Creating default startup script /home/student-00-621a4ce59b11/.vnc/xstartup
Starting applications specified in /home/student-00-621a4ce59b11/.vnc/xstartup
Log file is /home/student-00-621a4ce59b11/.vnc/instance-1:1.log

student-00-621a4ce59b11@instance-1:~$
```

6. If everything went fine your VNC server is now running and listening on port 5901. You can verify this with netcat from the Google Compute Engine instance:

```
$ nc localhost 5901
```

If the VNC server is working you will get the following output

RFB 003.008

```
student-00-621a4ce59b11@instance-1:~$ nc localhost 5901
RFB 003.008
```

7. Install VNC client. There are many options available, one of them is **RealVNC Viewer**. Install one but don't try to connect to your server just yet: it will fail as the firewall rules don't allow it. The download link for **RealVNC Viewer**

https://www.realvnc.com/en/connect/download/viewer/

8. In order to communicate with our instance we need its external IP. You can find it on the Developers Console for the virtual machine instance you created. Let's try to connect to it using netcat again:

$ nc 104.154.109.14 5901

Alternatively, you can use telnet

$ telnet 104.154.109.14 5901

Regardless of the tool you use the connection will fail, this is expected as the firewall rules block all communications by default for security reasons.

9. Let's fix that. Go to network details on the instance menu

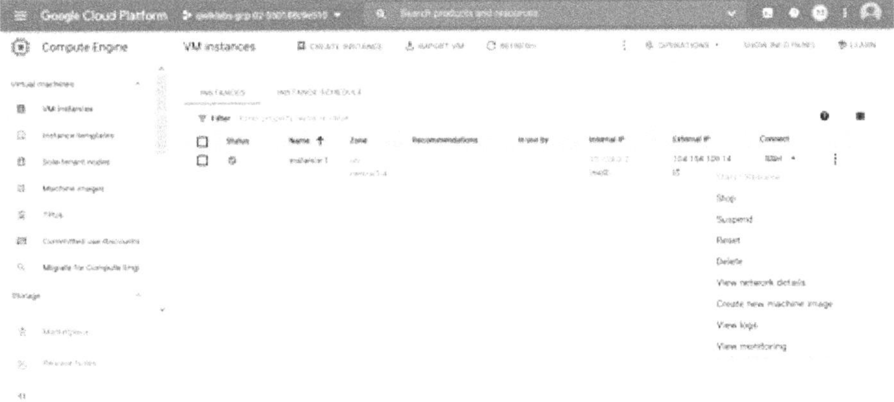

- Then choose networks/default https://console.cloud.google.com/networking/networks/details/default?

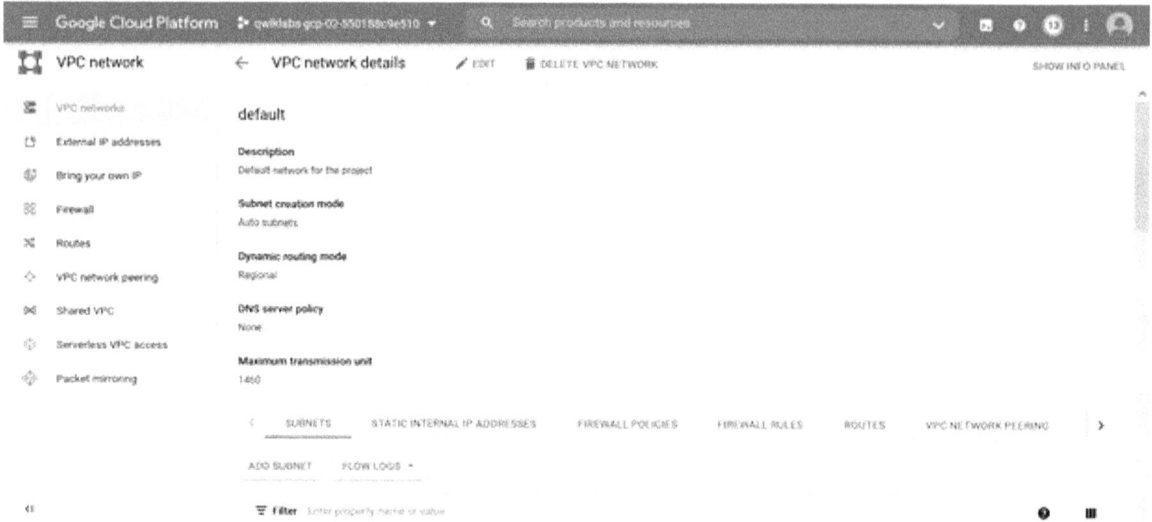

- Then choose Firewall Rules tab where you will see the list of firewall rules setup for your machine

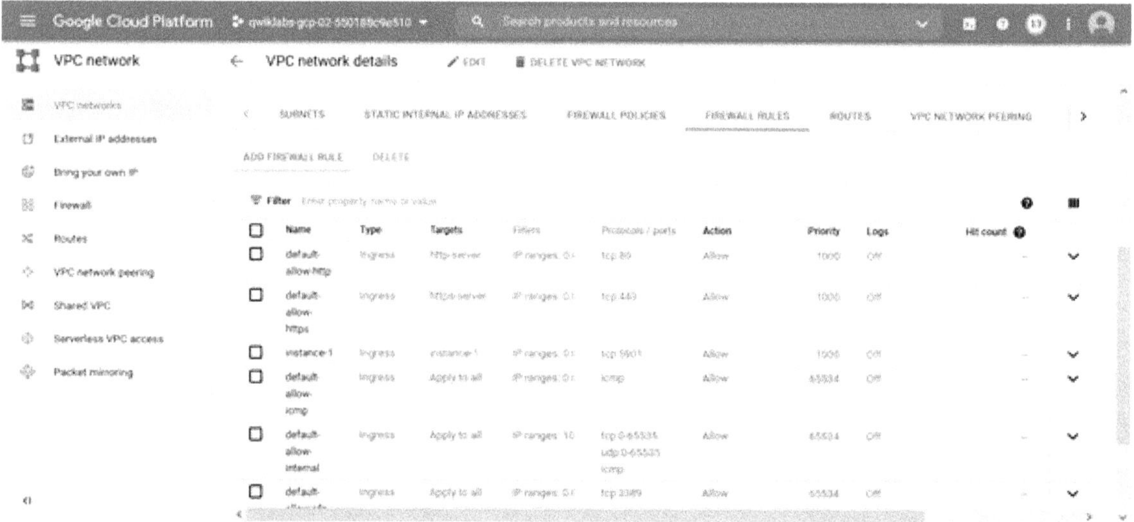

- Then choose to add new firewall rule

https://console.cloud.google.com/networking/firewalls/add

10. I created new firewall rule to allow incoming Tcp connection to the VNC server port 5901. The traffic will be on the port 5901 for protocol TCP and going to instances tagged as instance-1.

- You must setup the following details when creating the fire wall rule:
- In Name field: Choose a descriptive name for the rule. In my case I put the name “instance-1”

- In Source IP Ranges field: We will allow traffic coming from any source, which is why we use 0.0.0.0/0, the IP mask equivalent to a wildcard.
- In Allowed protocols or ports field: The traffic will be on the port 5901 for protocol TCP.
- In target tags field: The traffic will be going to instances tagged as "". In my case I tagged my instance as "instance-1" when I created the server. You can leave this field empty if you don't want to specify certain server in the network and you want to apply the rule on all servers in the network.

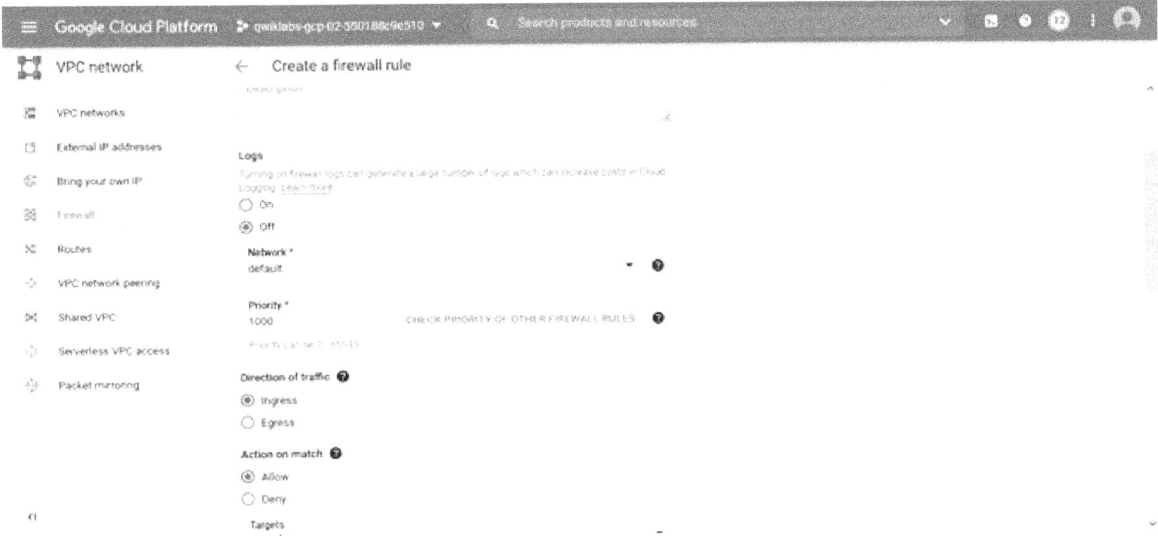

11. We must reset the server and start the vncserver after adding the firewall rule. Check that you have access to the VNC port 5901

$ sudo reboot

$ vncserver

$ nc 104.154.109.14 5901

```
student-00-621a4ce59b11@instance-1:~$ vncserver

New 'X' desktop is instance-1:1

Starting applications specified in /home/student-00-621a4ce59b11/.vnc/xstartup
Log file is /home/student-00-621a4ce59b11/.vnc/instance-1:1.log

student-00-621a4ce59b11@instance-1:~$ nc 104.154.109.14 5901
```

The connection is now allowed by the firewall by getting the following response.

$ nc 104.197.91.140 5901

RFB 003.008

6. Installing XRDP server:

1. Login to Google Cloud console. Navigate the left menu to list the VM instances running on your project: Compute > Compute Engine > VM instances. Click on the Create Instance button to access the instance creation form. Repeat the steps in section a to create Linux virtual machine

2. Connect to the Linux virtual machine using the Google Cloud browser. You will be connected to the machine using the SSH key you defined during machine setup, or the default SSH keys provided by Google Cloud.

3. Update the Linux machine packages using the command "sudo apt-get update"command.

```
$ sudo apt-get update

$ sudo apt-get upgrade
```

4. Install XRDP server

```
$ sudo apt-get install -y xrdp

$ sudo ufw allow 3389/tcp
```

5. Start XRDP server

```
$ sudo service xrdp restart
```

6. Usually there is default firewall rule that allows the RDP incoming connections to the server and so no need to add firewall rule to allow incoming RDP connections to port 3389.

7. You can check the RDP service is working by netcat command: You can verify this with netcat or telnet commands from the Google Compute Engine instance:

```
$ telnet 127.0.0.1 3389
```

If the RDP server is working you will get the following output

```
student-00-a3ac361471ed@instance-1:~$ telnet 127.0.0.1 3389
Trying 127.0.0.1...
Connected to 127.0.0.1.
Escape character is '^]'.
```

Use ctrl+] , then press q keys to exit

8. Check that you have access to the RDP port 3389 from your external network to the public IP of the server. In my example was:

$ telnet 34.132.152.62 3389

```
student-00-a3ac361471ed@instance-1:~$ telnet 34.132.152.62  3389
Trying 34.132.152.62...
Connected to 34.132.152.62.
Escape character is '^]'.
^]
telnet> q
Connection closed.
student-00-a3ac361471ed@instance-1:~$
```

9. When you connect to the Linux server through RDP connection, better to add a user and reset the root password so you connect to server by either root account or user account.

- Create username, as example hasooly using add user command. Specify the password and user information.

$ sudo adduser hasooly

```
student-00-a3ac361471ed@instance-1:~$ sudo adduser hasooly
Adding user `hasooly' ...
Adding new group `hasooly' (1003) ...
Adding new user `hasooly' (1002) with group `hasooly' ...
Creating home directory `/home/hasooly' ...
Copying files from `/etc/skel' ...
Enter new UNIX password:
Retype new UNIX password:
passwd: password updated successfully
Changing the user information for hasooly
Enter the new value, or press ENTER for the default
        Full Name []: hidaia
        Room Number []:
        Work Phone []:
        Home Phone []:
        Other []:
Is the information correct? [Y/n] y
student-00-a3ac361471ed@instance-1:~$
```

- To update also the root user, connect to the server using super root account. Then change the password of the root user by passwd command

$ sudo su

passwd

Give the new password of the root

```
root@instance-1:/home/student-00-a3ac361471ed# passwd
Enter new UNIX password:
Retype new UNIX password:
passwd: password updated successfully
root@instance-1:/home/student-00-a3ac361471ed#
```

7. Installing a Graphical User Interface (GUI) for Linux Google Cloud instance and connecting to the server through VNC or RDP connection:

I will show here installing graphical user interface (GUI) for Google Compute Engine instance and connecting to the server using VNC or RDP connection

There are many methods for installation of desktop environment in the Linux machine. I will mention two packages that can be used to access our instance through a desktop environment

a) Gnome: Gnome is beautiful … and heavy! So be patient while everything gets installed.

b) Xfce: If you prefer something faster to install you might like XFCE:

a) Installing XFCE as Graphical User Interface (GUI) and connecting to the Linux Machine instance through RDP and VNC connection:

1. Fellow the steps in the previous section to create Linux machine instance and XRDP server. This section is based on the setup done in the previous sections.

2. Install XRDP server

```
$ sudo apt-get  update

$ sudo apt-get  upgrade

$ sudo apt-get install -y xrdp
```

3. Start XRDP server

```
$ sudo service xrdp restart

$ sudo ufw allow 3389/tcp
```

4.Install XFCE packages

```
$ sudo apt-get install xfce4 xfce4-goodies

$ sudo apt-get install xfce4-terminal
```

5. Install Firefox internet browser

```
$ sudo apt install firebox
```

6. If you prefer Google chrome browser, you can download chromium-browser through the following command:

```
$ sudo apt install chromium-browser
```

7. When you connect to the Linux server through RDP connection, better to add a user and reset the root password so you connect to server by either root account or user account.

- Create username, as example hasooly using add user command. Specify the password and user information.

```
$ sudo adduser hasooly
```

```
student-00-a3ac361471ed@instance-1:~$ sudo adduser hasooly
Adding user `hasooly' ...
Adding new group `hasooly' (1003) ...
Adding new user `hasooly' (1002) with group `hasooly' ...
Creating home directory `/home/hasooly' ...
Copying files from `/etc/skel' ...
Enter new UNIX password:
Retype new UNIX password:
passwd: password updated successfully
Changing the user information for hasooly
Enter the new value, or press ENTER for the default
        Full Name []: hidaia
        Room Number []:
        Work Phone []:
        Home Phone []:
        Other []:
Is the information correct? [Y/n] y
student-00-a3ac361471ed@instance-1:~$
```

- To update also the root user, connect to the server using super root account. Then change the password of the root user by passwd command

  $ sudo su

  # passwd

  Give the new password of the root

```
root@instance-1:/home/student-00-a3ac361471ed# passwd
Enter new UNIX password:
Retype new UNIX password:
passwd: password updated successfully
root@instance-1:/home/student-00-a3ac361471ed#
```

8. Using windows remote access client, connect to the server by giving the IP address and the username and password in the server. In my case

   IP: (Provide public IP of the instance)

   Username: hasooly

   Password:

You will be logged on to the machine.

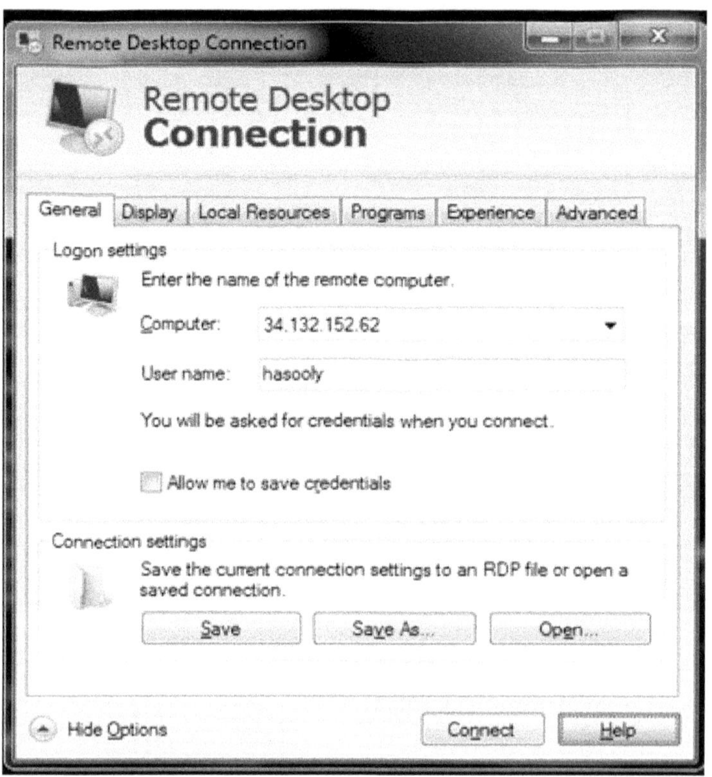

9. In case you need a VNC server to interact with desktop environment. You can use vnc4server, you can install your favorite one:

```
$ sudo apt-get install vnc4server
```

Or you can tightvncserver

```
$ sudo apt install -y tightvncserver
```

I used here tightvncserver

$ sudo apt install -y tightvncserver
$ sudo ufw allow 5901:5910/tcp

10 Start the VNC server, you'll then be prompted to create and verify a new password:

$ tightvncserver

```
student-00-621a4ce59b11@instance-1:~$ vncserver

You will require a password to access your desktops.

Password:
Verify:
xauth:  file /home/student-00-621a4ce59b11/.Xauthority does not exist

New 'X' desktop is instance-1:1

Creating default startup script /home/student-00-621a4ce59b11/.vnc/xstartup
Starting applications specified in /home/student-00-621a4ce59b11/.vnc/xstartup
Log file is /home/student-00-621a4ce59b11/.vnc/instance-1:1.log
```

11. If everything went fine your VNC server is now running and listening on port 5901. You can verify this with netcat from the Google Compute Engine instance:

$ nc localhost 5901
RFB 003.008

12. Make sure to create firewall rule to allow incoming Tcp connection to the VNC server port 5901 following the steps in previous sections.

13. We now need to kill the session we just created and make a tweak to the startup script for VNC Server to make it work properly. If we don't perform this step then all we will see is a grey cross-hatched screen with an "X" cursor and/or a grey screen with a Terminal Session, depending on the Ubuntu version. So, type the following command to kill the session:

$ vncserver -kill :1

14. A startup script is automatically generated by the shell, as shown in the following screenshot. This startup script can be accessed and edited by copying and pasting its `PATH` in the following manner: In my case I got the following xtratup script and log file.

Starting applications specified in /home/student-00-621a4ce59b11/.vnc/xstartup
Log file is /home/student-00-621a4ce59b11/.vnc/instance-1:1.log

Now open the file we need to edit:

$ vim .vnc/xstartup

The xstartup file will look when opened first time

```
#!/bin/sh

xrdb $HOME/.Xresources
xsetroot -solid grey
#x-terminal-emulator -geometry 80x24+10+10 -ls -title "$VNCDESKTOP Desktop" &
#x-window-manager &
# Fix to make GNOME work
export XKL_XMODMAP_DISABLE=1
/etc/X11/Xsession
```

15. Press the [Insert] key ("i" in Ubuntu) once (this will switch us into "edit" mode) and then edit the script so it ends up looking like this:

```
#!/bin/bash

xrdb $HOME/.Xresources

startxfce4 &
```

When you're done editing the .vnc/xstartup file for your particular version of Ubuntu press the [Esc] key once and type the following to save the changes and bring you back to the command line:

:wq

16. You must kill the vncservor, then start the server again

```
$ vncserver -kill :1
$  vncserver -geometry 1024x640
```

17. Install VNC client in your computer. There are many options available, one of them is RealVNC Viewer. Install one but don't try to connect to your server just yet: it will fail as the firewall rules don't allow it. The download link for RealVNC Viewer

https://www.realvnc.com/en/connect/download/viewer/

18. Open your VNC viewer and connect to the IP of your Compute Engine instance on port 5901.

19. Your Desktop environment is working. You can use Firefox browser for internet.

b) Installing GNOME as Graphical User Interface (GUI) and connecting to the Linux Machine instance through RDP connection:

1. Fellow the steps in the previous section to create Linux machine instance and XRDP server. This section is based on the setup done in the previous sections.

2. Install XRDP server

```
$ sudo apt-get  update

$ sudo apt-get  upgrade

$ sudo apt-get install -y xrdp

$ sudo ufw allow 3389/tcp
```

3. Start XRDP server

```
$ sudo service xrdp restart
```

4. Install GNOME packages through the following commands

```
$ sudo apt-get install gnome-shell
$ sudo apt-get install Ubuntu-gnome-desktop
$ sudo apt-get install autocutsel
$ sudo apt-get install gnome-core
$ sudo apt-get install gnome-panel
$ sudo apt-get install gnome-themes-standard
$ sudo apt-get install gnome-settings-daemon
$ sudo apt-get install metacity
$ sudo apt-get install gnome-terminal
```

5. Install firefox internet browser

```
$ sudo apt install firefox
```

6. If you prefer Google chrome browser, you can download chromium-browser through the following command:

```
$ sudo apt install chromium-browser
```

7. When you connect to the Linux server through RDP connection, better to add a user and reset the root password so you connect to server by either root account or user account.

- Create username, as example hasooly using add user command. Specify the password and user information.

$ sudo adduser hasooly

```
student-00-a3ac361471ed@instance-1:~$ sudo adduser hasooly
Adding user `hasooly' ...
Adding new group `hasooly' (1003) ...
Adding new user `hasooly' (1002) with group `hasooly' ...
Creating home directory `/home/hasooly' ...
Copying files from `/etc/skel' ...
Enter new UNIX password:
Retype new UNIX password:
passwd: password updated successfully
Changing the user information for hasooly
Enter the new value, or press ENTER for the default
        Full Name []: hidaia
        Room Number []:
        Work Phone []:
        Home Phone []:
        Other []:
Is the information correct? [Y/n] y
student-00-a3ac361471ed@instance-1:~$
```

- To update also the root user, connect to the server using super root account. Then change the password of the root user by passwd command

  $ sudo su

  # passwd

  Give the new password of the root

```
root@instance-1:/home/student-00-a3ac361471ed# passwd
Enter new UNIX password:
Retype new UNIX password:
passwd: password updated successfully
root@instance-1:/home/student-00-a3ac361471ed#
```

8. Using windows remote access client, connect to the server by giving the IP address and the username and password in the server. In my case

IP: (Provide public IP of the instance)

Username: hasooly

Password:

You will be logged on to the machine.

Remote Desktop Connection
Remote Desktop
Connection
General
Display
Local Resources
Programs
Experience
Advanced
Logon settings
Enter the name of the remote computer.
Computer:
34.132.152.62
User name:
hasooly
You will be asked for credentials when you connect.
Allow me to save credentials
Connection settings
Save the current connection settings to an RDP file or open a saved connection.
Save
Save As...
Open...
Hide Options
Connect
Help

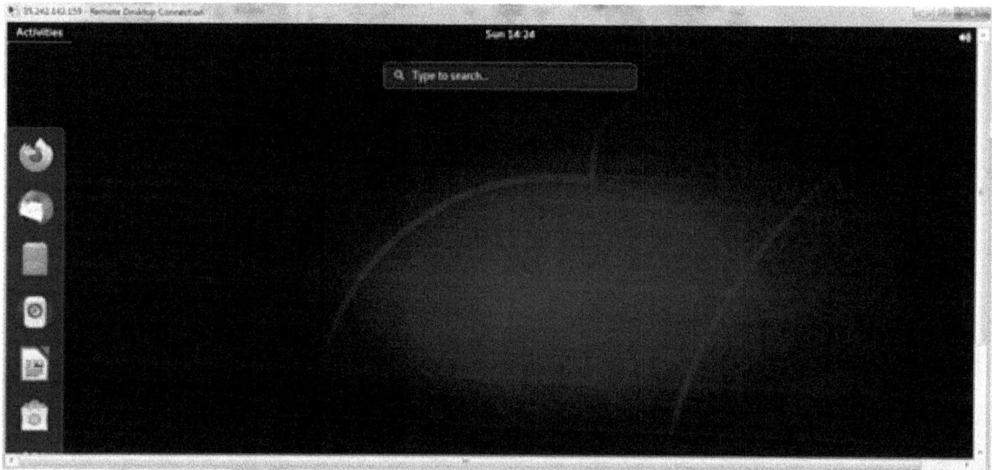
Activities
Type to search...

c) Installing GNOME as Graphical User Interface (GUI) and connecting to the Linux Machine instance through VNC connection:

1. Fellow the steps in the previous section to create Linux machine instance and VNC server. This section is based on the setup done in the previous sections.

2. VNC server requires some packages to load GUI. These packages are: install Ubuntu-desktop, gnome-panel, gnome-settings, -daemon, metacity, nautlius, gnome-terminal.

3.Install GNOME packages through the following commands

```
$ sudo apt-get update
$ sudo apt-get upgrade
$ sudo apt-get install gnome-shell
$ sudo apt-get install ubuntu-gnome-desktop
$ sudo apt-get install autocutsel
$ sudo apt-get install gnome-core
$ sudo apt-get install gnome-panel
$ sudo apt-get install gnome-themes-standard
$ sudo apt-get clean
```

4. Then we install VNC4server. And using the VNC client in windows machine, we connect to graphical user interface if Ubuntu Server instance in Google Cloud Platform.

```
$ sudo apt-get install vnc4server

$ touch ~/.Xresources
```

Or you can install tightvncserver

```
$ sudo apt-get install tightvncserver
$ touch ~/.Xresources
```

In this example I used tightvncserver

5. You can clean uninstalled packages to free space

```
$ sudo apt-get clean
```

6. After installing vnc4server, launch the server by typing the following command: Start the VNC Server, you'll then be prompted to create and verify a new password:

```
$ vncserver
```

Or in case of tightvncserver

$ tightvncserver

```
cloudsigma@UP-5586: ~
cloudsigma@UP-5586:~$ tightvncserver

You will require a password to access your desktops.

Password:
Verify:
Would you like to enter a view-only password (y/n)? y
Password:
Verify:
xauth:  file /home/cloudsigma/.Xauthority does not exist

New 'X' desktop is UP-5586:1

Creating default startup script /home/cloudsigma/.vnc/xstartup
Starting applications specified in /home/cloudsigma/.vnc/xstartup
Log file is /home/cloudsigma/.vnc/UP-5586:1.log

cloudsigma@UP-5586:~$
```

7. Allow the incoming connections to port 5901 by creating firewall rule that allows all incoming connection to port 5901 on the server using the steps mentioned in previous section. You can also try this command

$ sudo ufw allow 5901:5910/tcp

8. If everything went fine your VNC server is now running and listening on port 5901. You can verify this with netcat applied to the public IP of the server

$ nc (Public IP Address) 5901
RFB 003.008

9. Open your VNC viewer and connect to the IP of your Compute Engine instance on port 5901. You will be connected to the command line console of the server.

10. You need to specify a startup script. The startup script will allow us to access GNOME panel using the VNC connection. We need to modify the startup script. vnc/startup. Edit the default startup scrip. We are going to define gnome-panel gnome-settings-daemon metacity nautilus gnome-terminal. Type i to enter INSERT mode. Next, delete all the text in the startup script. Copy paste the following code into the startup script:

$ vim .vnc/xstartup

((

#!/bin/sh

autocutsel -fork

xrdb $HOME/.Xresources

xsetroot -solid grey

export XKL_XMODMAP_DISABLE=1

export XDG_CURRENT_DESKTOP="GNOME-Flashback:Unity"

export XDG_MENU_PREFIX="gnome-flashback-"

unset DBUS_SESSION_BUS_ADDRESS

gnome-session --session=gnome-flashback-metacity --disable-acceleration-check --debug &

((

11. You must kill the vncservor, then start the server again

    $ vncserver -kill :1
    $ vncserver -geometry 1024x640

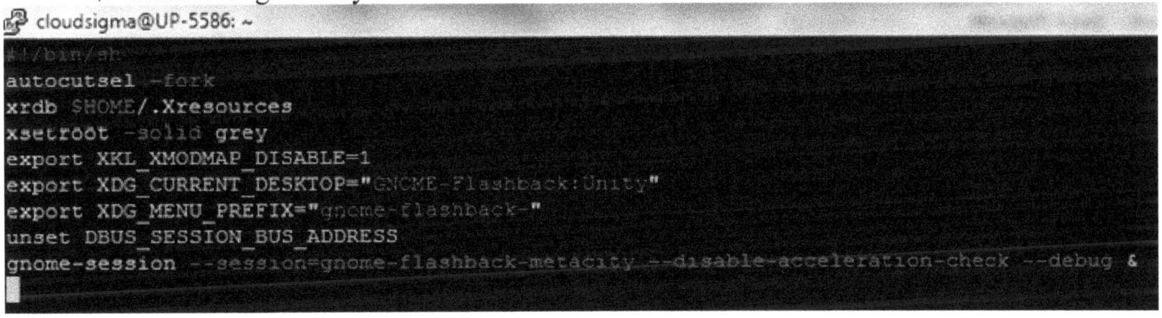

cloudsigma@UP-5586: ~

```
#!/bin/sh
autocutsel -fork
xrdb $HOME/.Xresources
xsetroot -solid grey
export XKL_XMODMAP_DISABLE=1
export XDG_CURRENT_DESKTOP="GNOME-Flashback:Unity"
export XDG_MENU_PREFIX="gnome-flashback-"
unset DBUS_SESSION_BUS_ADDRESS
gnome-session --session=gnome-flashback-metacity --disable-acceleration-check --debug &
```

12. Install firebox internet browser

    $ sudo apt install firebox –y

13. If you prefer Google chrome browser, you can download chromium-browser through the following command:

    $ sudo apt install chromium-browser

14. Open your VNC viewer and connect to the IP of your Compute Engine instance on port 5901. You will be connected to the Graphical User Interface GUI of the server.

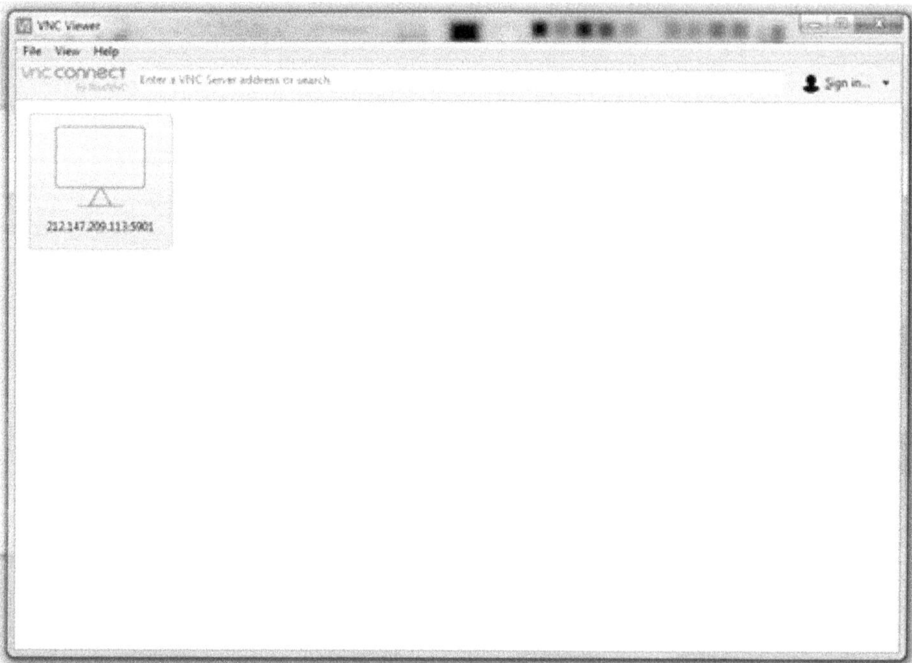

8. Quick guide to create a Linux virtual machine in Cloudsigma:

CloudSigma offers a range of locations from Europe to the United States, the Middle East and APAC. We are adding new locations over time as we expand our offering globally.

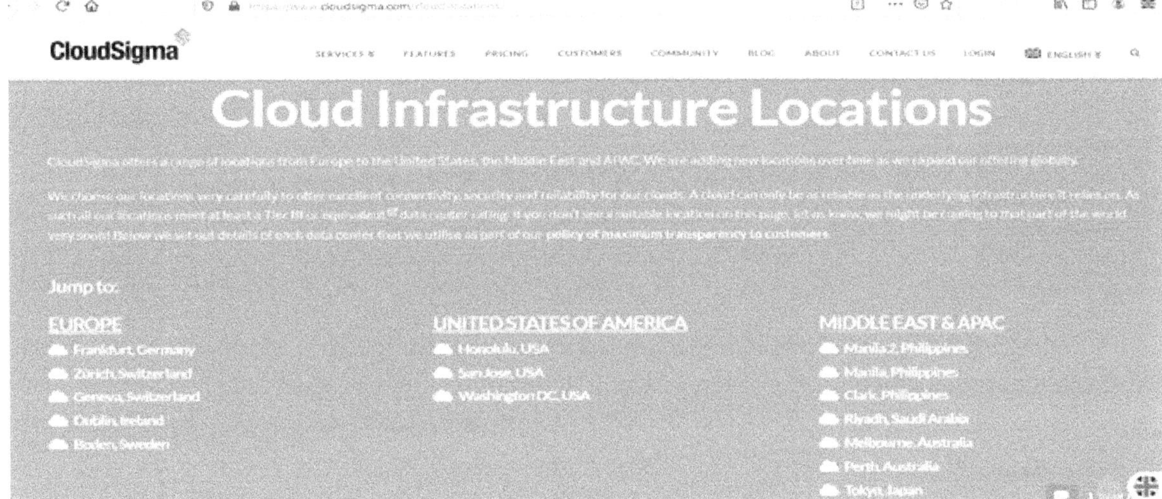

1. I chose to create free trial account at server in Ireland. So, I registered account under the link https://ec.servecentric.com/ui/4.0/login (You can change the server country from upper tab)

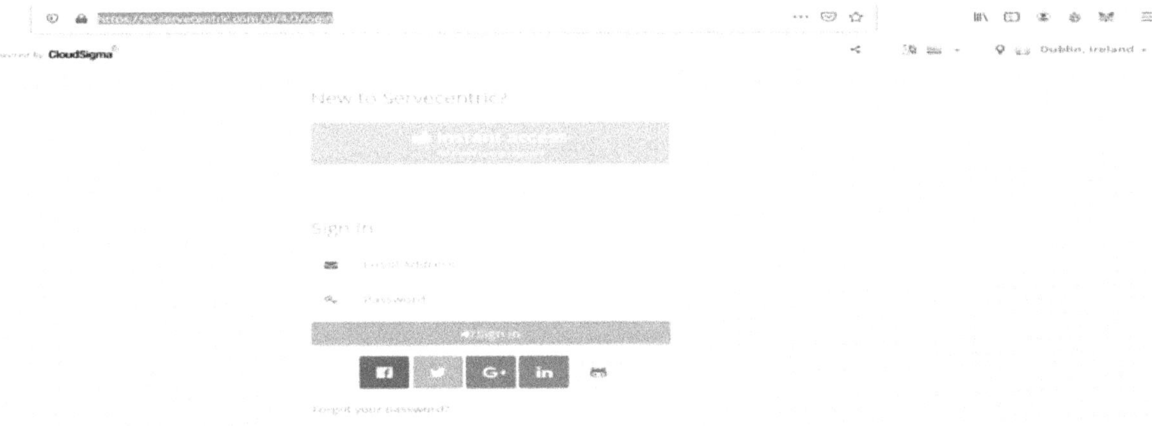

2. Choose "Instant access". You will be forwarded to the following dashboard:

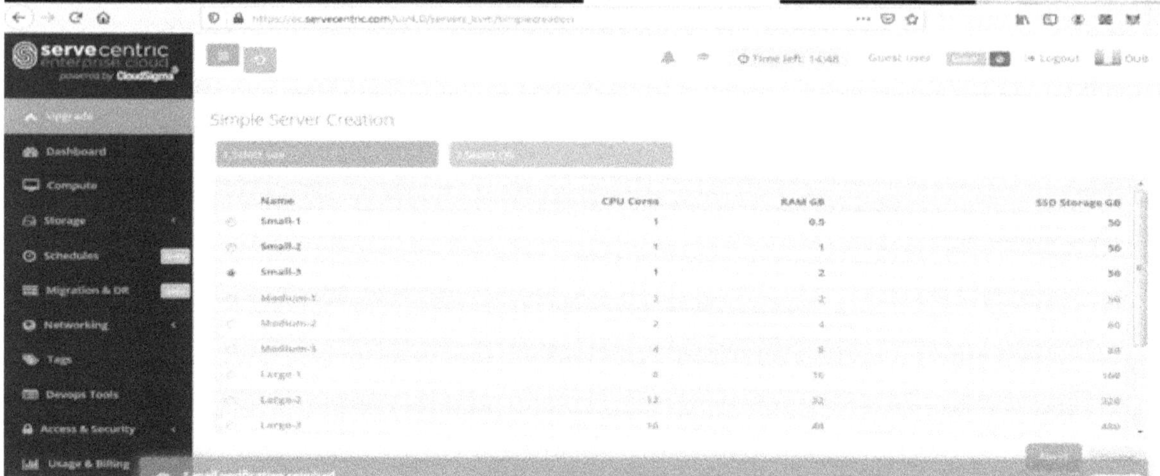

3. Choose upgrade and verify your phone number email. I registered with temporary email. When they accept to provide you trial account, you will get trial icon in the top of the page

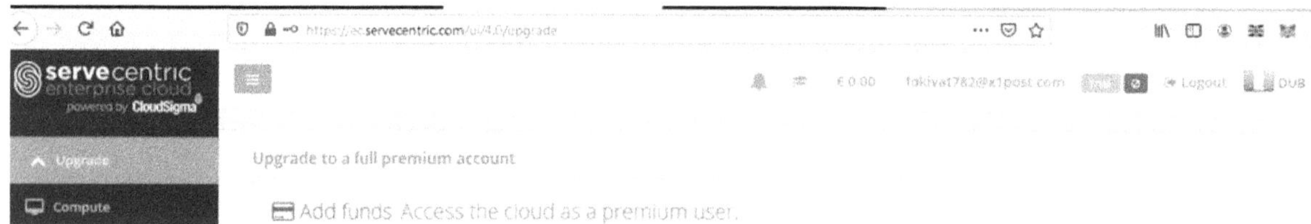

4. You upgrade by verifying the phone number. You must have business IP to be able to upgrade for 7 days free trial. I upgraded and got the following dashboard. A Linux server was created for my account.

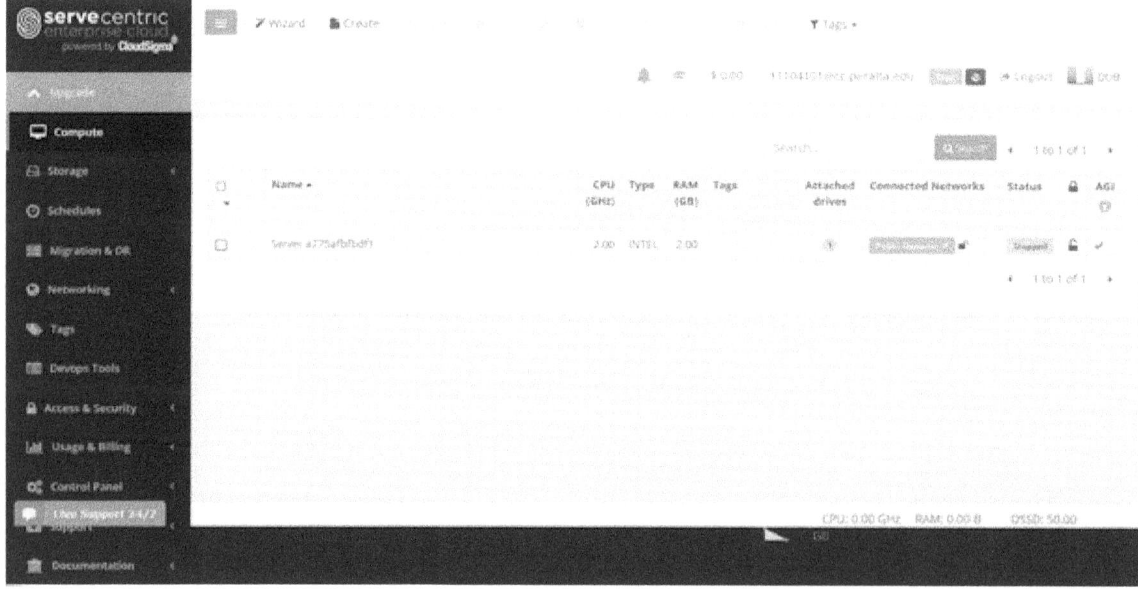

5. Delete the default server as you might have problem to login with the correct password.

6. Select Compute tab, then Wized tab, you will be provided with the list of servers that you can use within the country you registered under (Ireland in my case). Select server size. I chose server size small-3, 1 GB Ram, 50 GB Hard disks

7. Choose the server OS. I chose Ubuntu 18.04 LTS, as Windows not available in free trial. Then create the server.

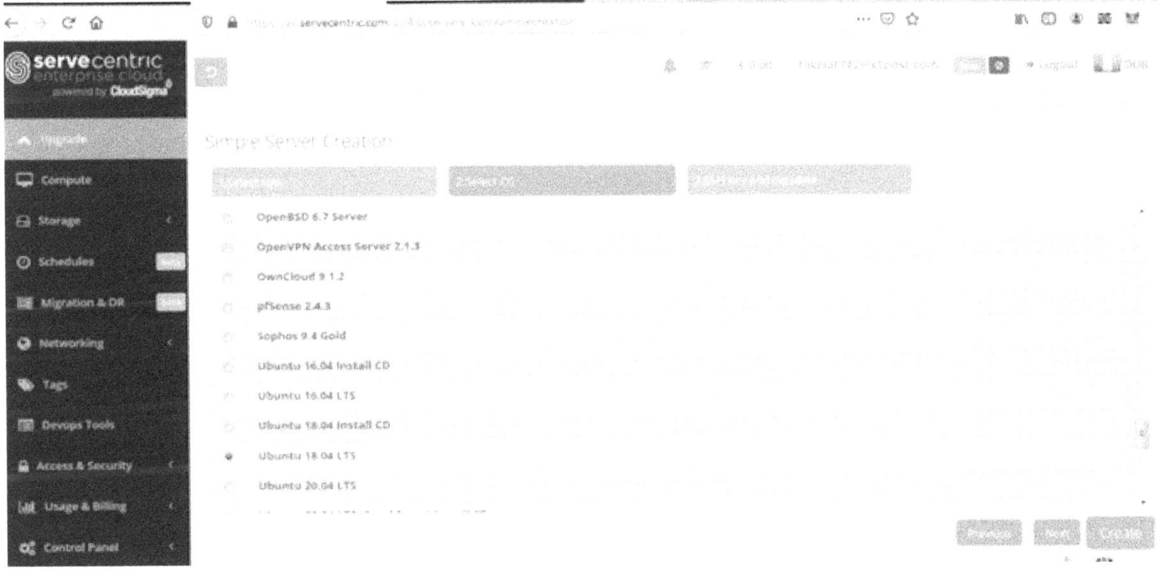

8. When the server created, you will get your credentials: IP address of Linux server, Username and Password.

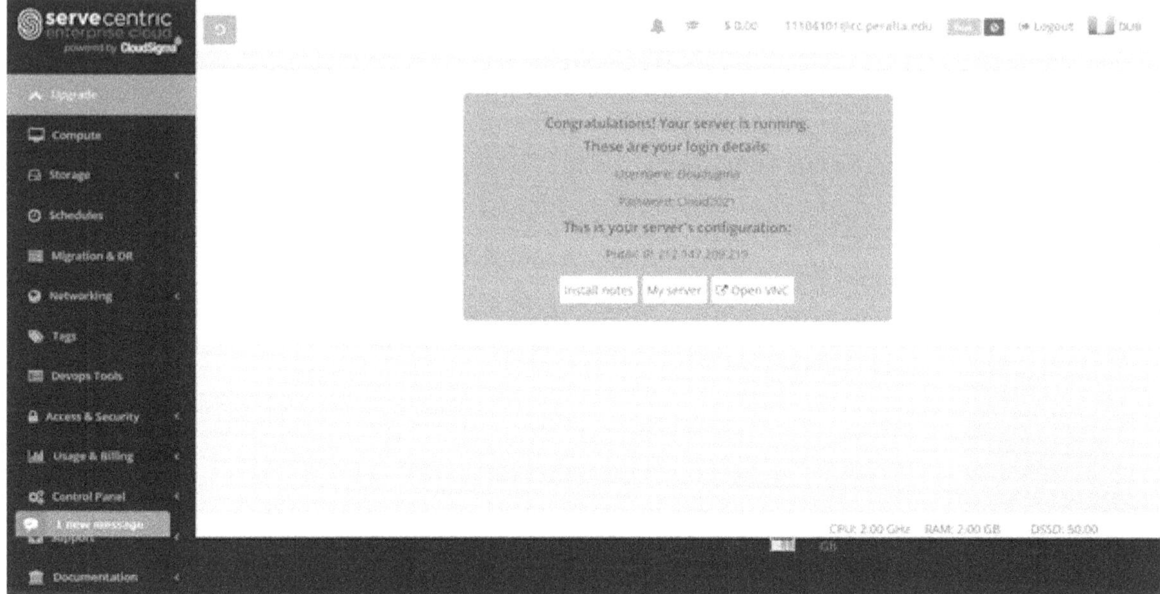

9. Click on the server, you will get the following sections for server details:

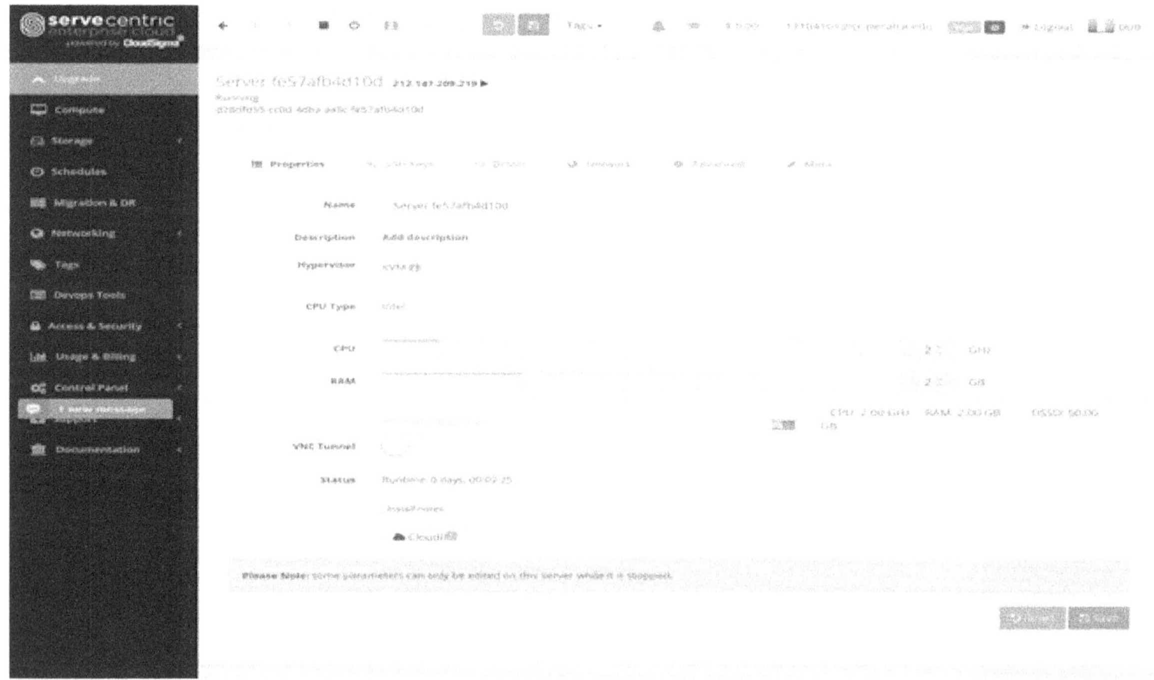

10. Choose the Drives tab to get information on the server type. In my case Ubuntu18.04 LTS Mounted on the server

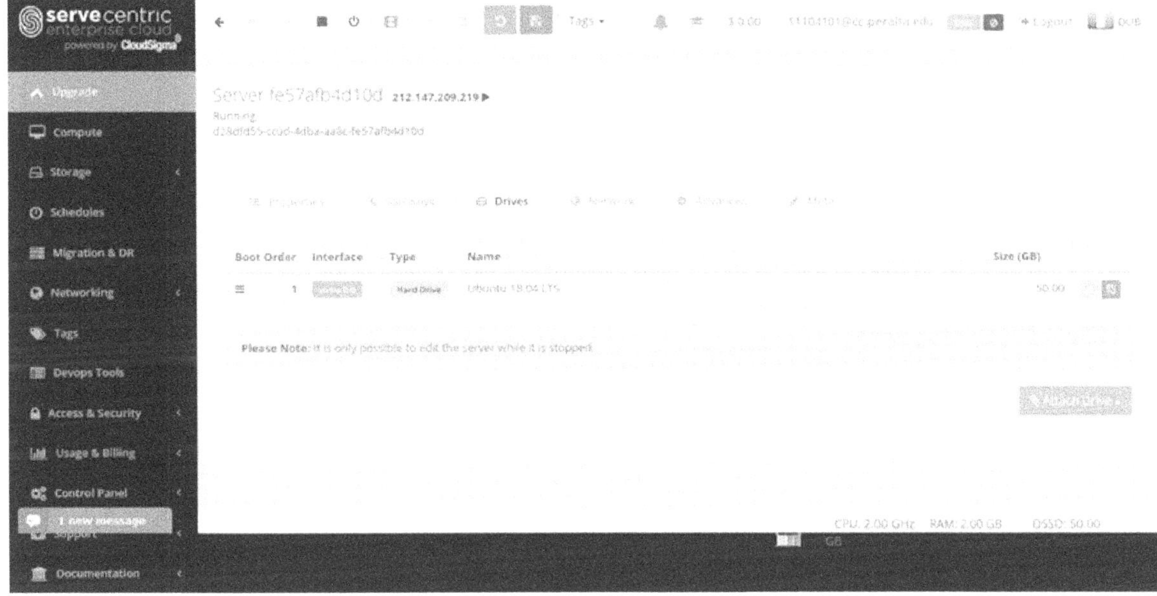

11. Click on Ubuntu 18.04 LTS

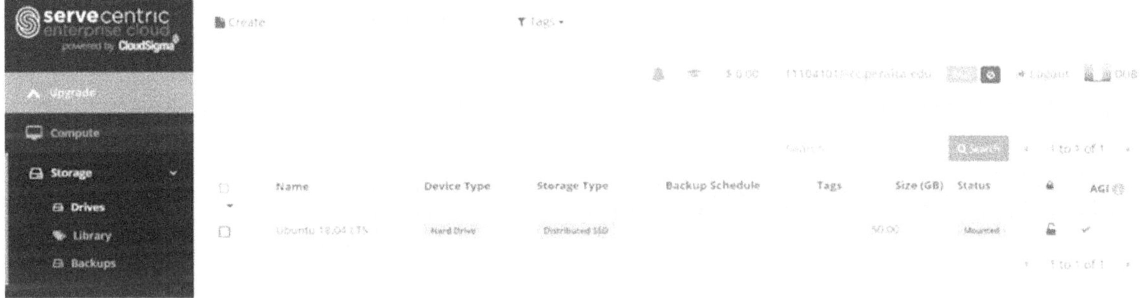

12. Go to install notes page. Here you will find the installation notes:

13. Initial Credentials: Username: cloudsigma Password: Cloud2021

14. Connect to your server via VNC. Go to the “Properties” tab of the server and Turn on the VNC Tunnel by clicking the toggle button labeled "VNC Tunnel". To use the web VNC client click on "View VNC"

15. You will be forwarded to the VNC command console. You will be requested to change your user password. So, your username will be cloudsigma, and password will be the new password you give.

16. Here screenshot of the VNC command line console. To check your username

$ whoami

```
cloudsigma@dsa:~$ whoami
cloudsigma
cloudsigma@dsa:~$
```

17. Use the PuTTY tool to connect to the server through the SSH connection and to install XRDP and XFCE package in the same steps as in the previous sections.

18. Install XRDP server

$ sudo apt-get update

$ sudo apt-get upgrade

$ sudo apt-get install -y xrdp

19. Start XRDP server

$ sudo service xrdp restart

$ sudo ufw allow 3389/tcp

20. Install XFCE packages

$ sudo apt-get install xfce4 xfce4-goodies -y

$ sudo apt-get install xfce4-terminal –y

21. Install firefox internet browser

$ sudo apt install firefox –y

22. If you prefer Google chrome browser, you can download chromium-browser through the following command:

$ sudo apt install chromium-browser

23. When you connect to the Linux server through RDP connection, better to add a user and reset the root password so you connect to server by either root account or user account.

- Create username, as example hasooly using add user command. Specify the password and user information.

$ sudo adduser hasooly

```
student-00-a3ac361471ed@instance-1:~$ sudo adduser hasooly
Adding user `hasooly' ...
Adding new group `hasooly' (1003) ...
Adding new user `hasooly' (1002) with group `hasooly' ...
Creating home directory `/home/hasooly' ...
Copying files from `/etc/skel' ...
Enter new UNIX password:
Retype new UNIX password:
passwd: password updated successfully
Changing the user information for hasooly
Enter the new value, or press ENTER for the default
        Full Name []: hidaia
        Room Number []:
        Work Phone []:
        Home Phone []:
        Other []:
Is the information correct? [Y/n] y
student-00-a3ac361471ed@instance-1:~$
```

- To update also the root user, connect to the server using super root account. Then change the password of the root user by passwd command

  $ sudo su

  # passwd

  Give the new password of the root

```
root@instance-1:/home/student-00-a3ac361471ed# passwd
Enter new UNIX password:
Retype new UNIX password:
passwd: password updated successfully
root@instance-1:/home/student-00-a3ac361471ed#
```

24. Using windows remote access client, connect to the server by giving the IP address and the username and password in the server. In my case

    IP: (Provide public IP of the instance)

    Username: hasooly

    Password:

You will be logged on to the machine.

25. In case you need a VNC server to interact with desktop environment. You can use vnc4server, you can install your favorite one:

```
$ sudo apt-get install vnc4server
```

Or you can tightvncserver

```
$ sudo apt install -y tightvncserver
```

I used here tightvncserver

```
$ sudo apt install -y tightvncserver
$ sudo ufw allow 5901:5910/tcp
```

26 Start the vnc server, you'll then be prompted to create and verify a new password:

$ tightvncserver

27. If everything went fine your VNC server is now running and listening on port 5901. You can verify this with netcat from the Google Compute Engine instance:

$ nc localhost 5901
RFB 003.008

28. Make sure to create firewall rule to allow incoming Tcp connection to the VNC server port 5901 following the steps in previous sections.

29. We now need to kill the session we just created and make a tweak to the startup script for VNC Server to make it work properly. If we don't perform this step then all we will see is a grey cross-hatched screen with an "X" cursor and/or a grey screen with a Terminal Session, depending on the Ubuntu version. So, type the following command to kill the session:

$ vncserver -kill :1

30. A startup script is automatically generated by the shell, as shown in the following screenshot. This startup script can be accessed and edited by copying and pasting its PATH in the following manner: In my case I got the following xtratup script and log file.

Starting applications specified in /home/student-00-621a4ce59b11/.vnc/xstartup
Log file is /home/student-00-621a4ce59b11/.vnc/instance-1:1.log

Now open the file we need to edit:

$ vim .vnc/xstartup

The xstartup file will look when opened first time

```
#!/bin/sh

xrdb $HOME/.Xresources
xsetroot -solid grey
#x-terminal-emulator -geometry 80x24+10+10 -ls -title "$VNCDESKTOP Desktop" &
#x-window-manager &
# Fix to make GNOME work
export XKL_XMODMAP_DISABLE=1
/etc/X11/Xsession
~
```

31. Press the [Insert] key ("i" in Ubuntu) once (this will switch us into "edit" mode) and then edit the script so it ends up looking like this:

```
#!/bin/bash

xrdb $HOME/.Xresources

startxfce4 &
```

When you're done editing the .vnc/xstartup file for your particular version of Ubuntu press the [Esc] key once and type the following to save the changes and bring you back to the command line:

:wq

32. You must kill the vncservor, then start the server again

```
$ vncserver -kill :1

$  vncserver
```

33. Install VNC client in your computer. There are many options available, one of them is **RealVNC Viewer**. Install one but don't try to connect to your server just yet: it will fail as the firewall rules don't allow it. The download link for **RealVNC Viewer**

https://www.realvnc.com/en/connect/download/viewer/

34. Open your VNC viewer and connect to the IP of your Compute Engine instance on port 5901.

35. Your Desktop environment is working. You can use Firefox browser for internet.

VNC Viewer
File View Help
Enter a VNC Server address or search
Sign in...
34.89.94.3:5901
34.89.94.3:5901 - VNC Viewer
Connecting to 34.89.94.3:5901...
Stop

9. Quick guide to create a Linux Virtual Machine in the Microsoft Azure portal:

a) Getting free Azure subscription through Sandbox Microsoft Learn subscription (No credit card needed):

A sandbox gives you access to Azure resources. Your Azure subscription will not be charged. The sandbox may only be used to complete training on Microsoft Learn. Use for any other reason is prohibited, and may result in permanent loss of access to the sandbox. The sandbox may only be used to complete training on Microsoft Learn.

You can obtain Sandbox free subscription for duration of 1or 2 hrs. each time through some exercises in Microsoft learn, as an example these exercises

- https://docs.microsoft.com/en-us/learn/modules/deploy-vms-from-vhd-templates/4-exercise-create-image-provision-vm?pivots=windows-cloud
- https://docs.microsoft.com/en-us/learn/modules/create-windows-virtual-machine-in-azure/3-exercise-create-a-vm
- https://docs.microsoft.com/en-us/learn/modules/create-linux-virtual-machine-in-azure/
- https://docs.microsoft.com/en-us/learn/modules/improve-app-scalability-resiliency-with-load-balancer/
- https://docs.microsoft.com/en-us/learn/modules/connect-on-premises-network-with-vpn-gateway/3-exercise-prepare-azure-and-on-premises-vnets-using-azure-cli-commands.

Creating the Azure Windows Virtual machine using sandbox which is learning subscription. You can connect to the VM via RDP port but you cannot access to Internet through the Internet Explorer.

Outbound data transfer is charged at the normal rate and inbound data transfer is free. Refer to these articles for more information:

https://docs.microsoft.com/en-us/learn/support/faq?pivots=sandbox
https://azure.microsoft.com/en-us/pricing/details/bandwidth/

b) Creating Linux Virtual Machine in the Azure portal:

1. Sign in to Azure

2. Create virtual machine

- Type **virtual machines** in the search.
- Under **Services**, select **Virtual machines**.
- In the **Virtual machines** page, select **Add**. The **Create a virtual machine** page opens.
- In the **Basics** tab, under **Project details**, make sure the correct subscription is selected and then choose to **Create new** resource group. In case of using Sandbox, Microsoft Learn subscription, a default resource group will be created.

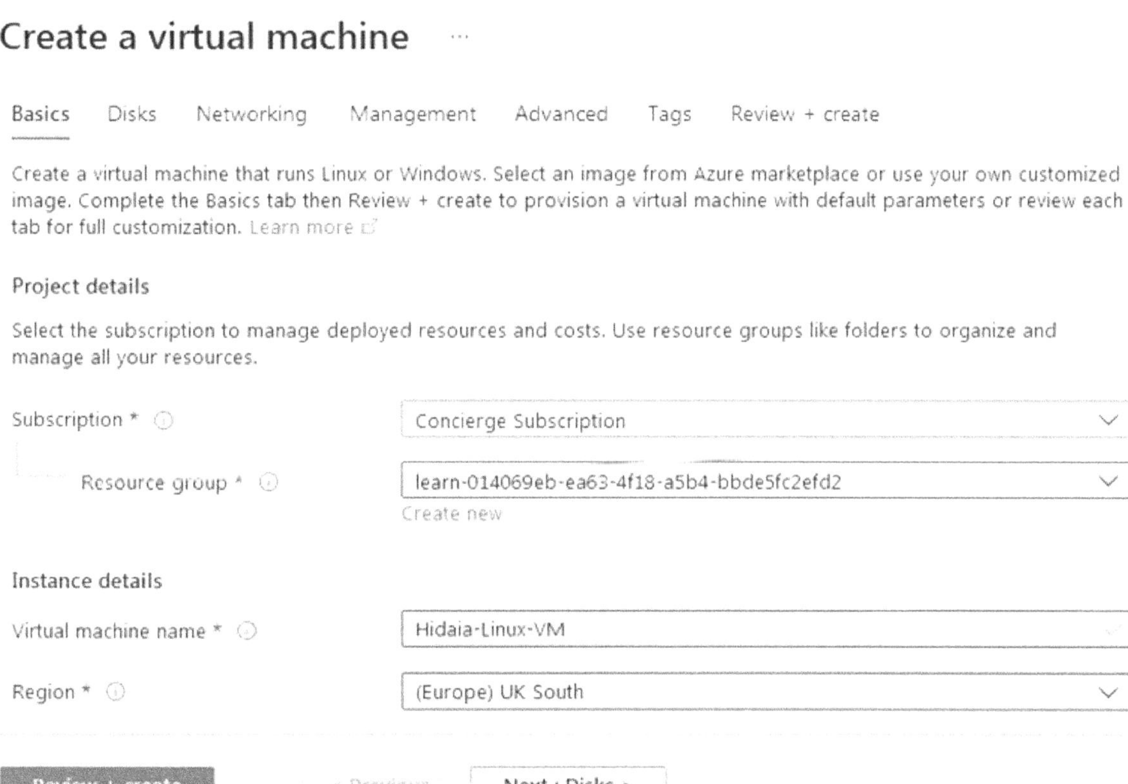

- Under Instance details, type the Virtual machine name, choose your preferred Region, and choose Linux machine image, such as Ubuntu 18.04 LTS. Leave the other defaults. In this example I setup my machine name "Hidaia-Linux-VM" and the region "South UK"

- Under Administrator account, you can select to login through SSH public key or through username and password.
- In case you want to login through SSH key, you can choose to enter your SSH key generated using PuTTY key generator, or azure can generate SSH key for you, or you can use existing azure SSH key. And enter also your username

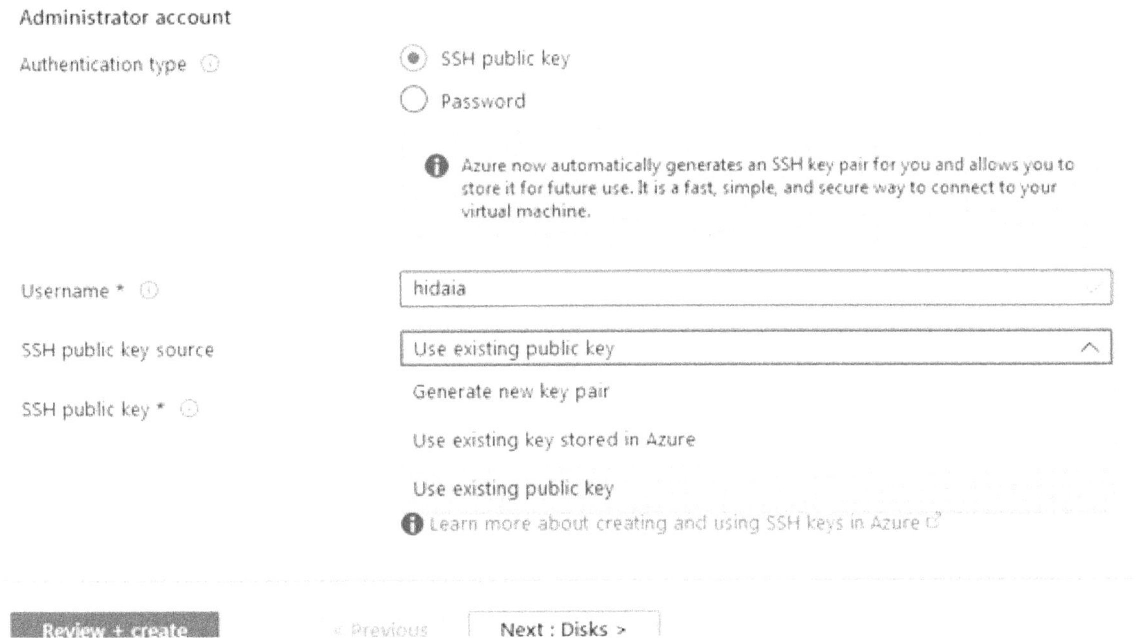

- In case you want to login through username and password, choose your username and password. In this example I chose to login by username and password.

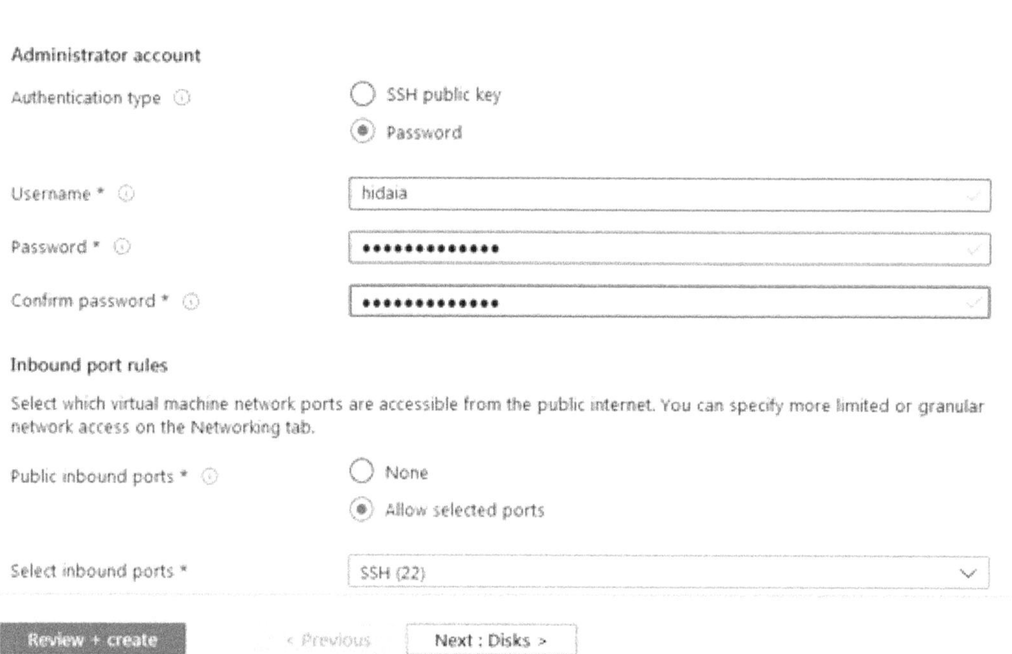

- Under Inbound port rules > Public inbound ports, choose Allow selected ports and then select SSH (22), HTTPS (443) and HTTP (80) from the drop-down.

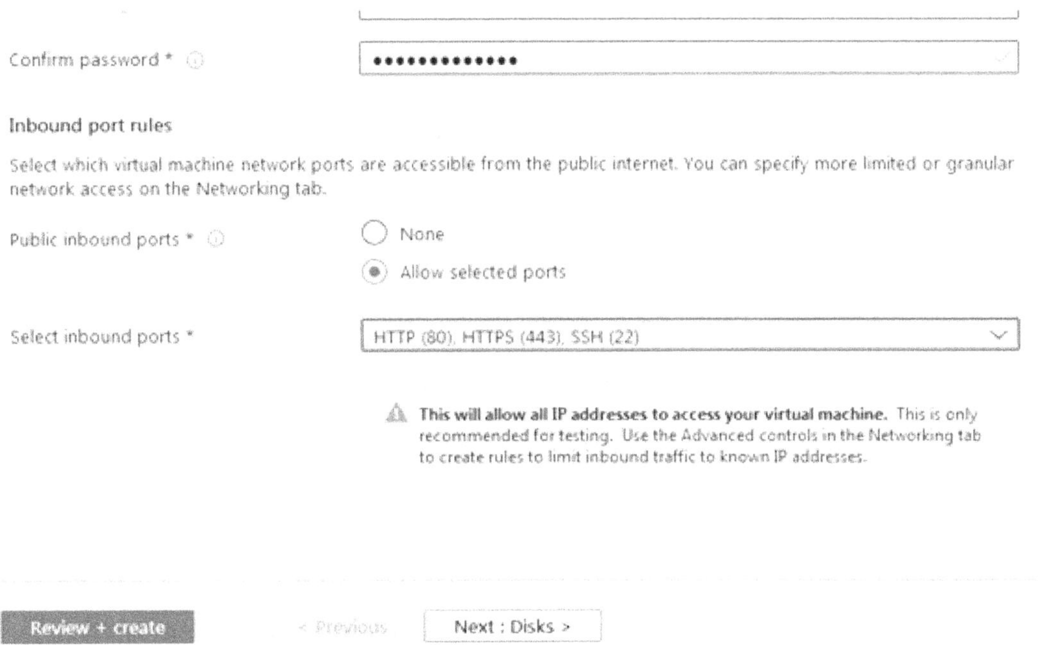

- Next page, you will be forwarded to configure the virtual machine Disk. Azure VMs have one operating system disk and a temporary disk for short-term storage. You can attach additional data disks. The size of the VM determines the type of storage you can use and the number of data disks allowed. In my example, I left the default configuration.

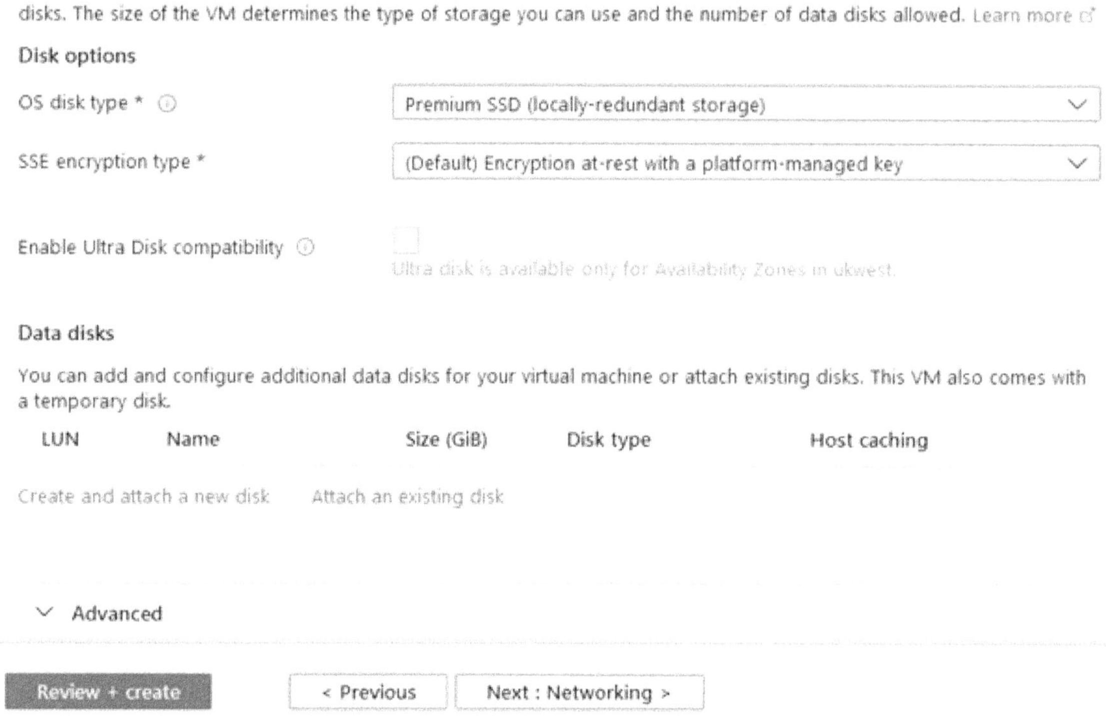

- In the next page, you will be forwarded to Define network connectivity for your virtual machine. Define network connectivity for your virtual machine by configuring network interface card (NIC) settings. You can control ports, inbound and outbound connectivity with security group rules, or place behind an existing load balancing solution. In my example, I chose the default configuration, but I added The RDP 3389 port among the inbound ports.

Home >

Create a virtual machine ...

Basics Disks **Networking** Management Advanced Tags Review + create

Define network connectivity for your virtual machine by configuring network interface card (NIC) settings. You can control ports, inbound and outbound connectivity with security group rules, or place behind an existing load balancing solution.
Learn more

Network interface

When creating a virtual machine, a network interface will be created for you.

Virtual network *	(new) learn-014069eb-ea63-4f18-a5b4-bbde5fc2efd2-vnet Create new
Subnet *	(new) default (10.0.0.0/24)
Public IP	(new) Hidaia-Linux-VM-ip Create new
NIC network security group	○ None ◉ Basic ○ Advanced

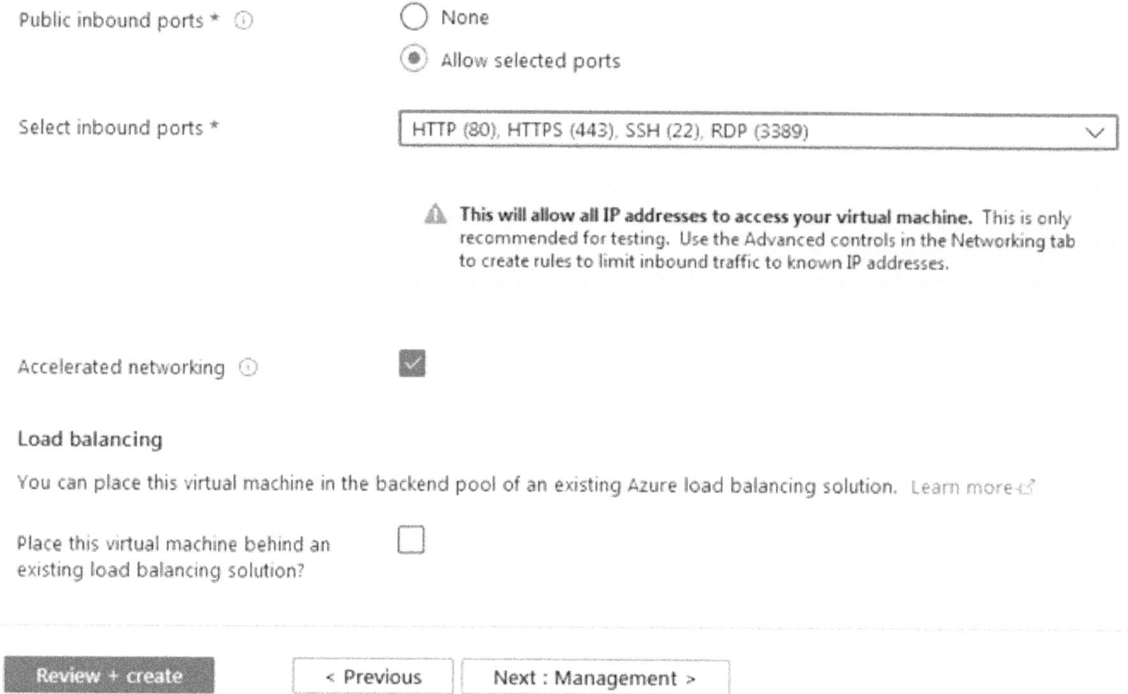

- In the next page configure monitoring and management options for your VM. Azure Security Center provides unified security management and advanced threat protection across hybrid cloud workloads. In this example, I left the default configuration.

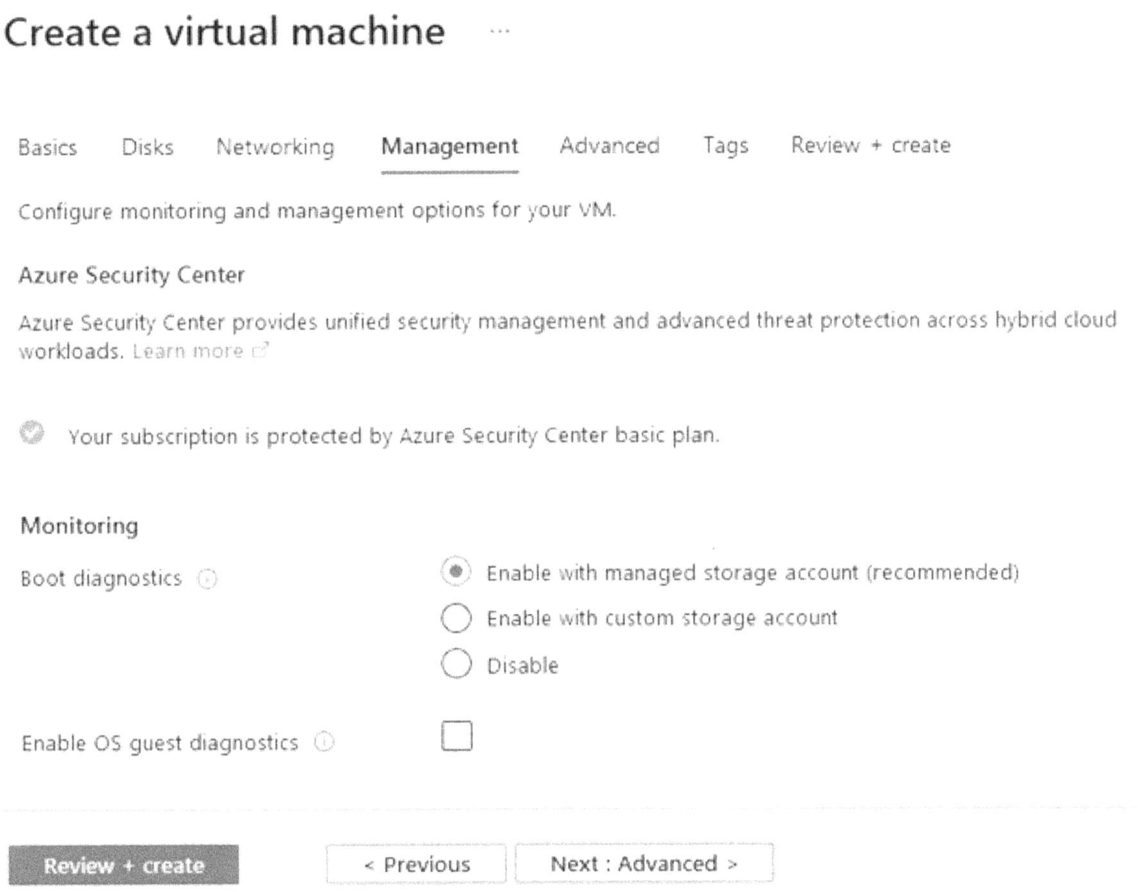

Create a virtual machine ...

Basics Disks Networking **Management** Advanced Tags Review + create

Configure monitoring and management options for your VM.

Azure Security Center

Azure Security Center provides unified security management and advanced threat protection across hybrid cloud workloads. Learn more

Your subscription is protected by Azure Security Center basic plan.

Monitoring

Boot diagnostics

(•) Enable with managed storage account (recommended)

( ) Enable with custom storage account

( ) Disable

Enable OS guest diagnostics [ ]

Review + create < Previous Next : Advanced >

- In the next page, Add additional configuration, agents, scripts or applications via virtual machine extensions or cloud-init. In this example, I left the default configuration.
- In the next page, add your Linux machine tags. Tags are name/value pairs that enable you to categorize resources and view consolidated billing by applying the same tag to multiple resources and resource groups. In this example, I left the default configuration.
- Leave the remaining defaults and then select the Review + create button at the bottom of the page. If you requested to download private SSH key, download the SSH key and create resource.

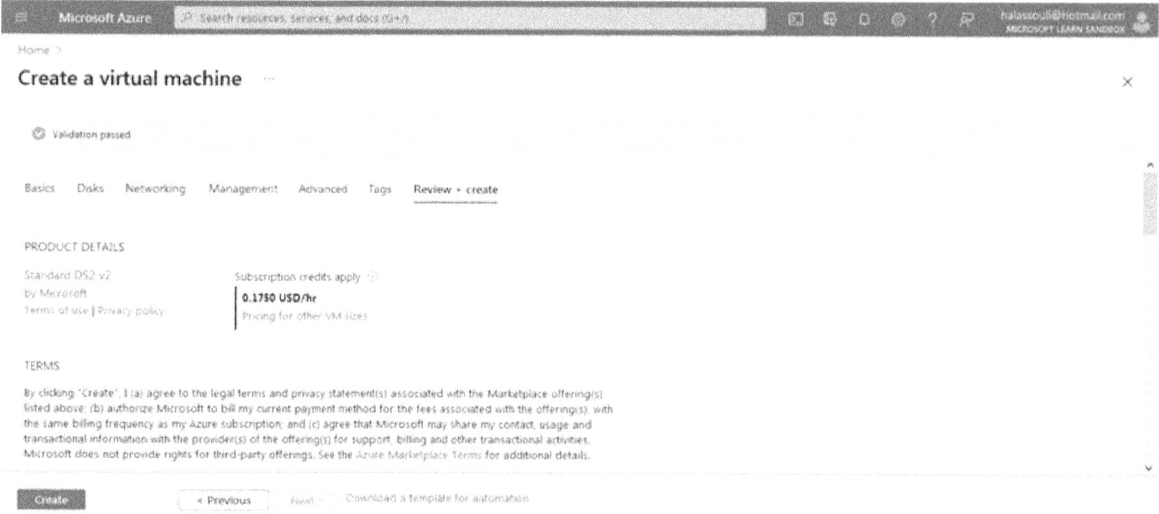

3. When the deployment is finished, select Go to resource. You will see the following dashboard of the new VM.

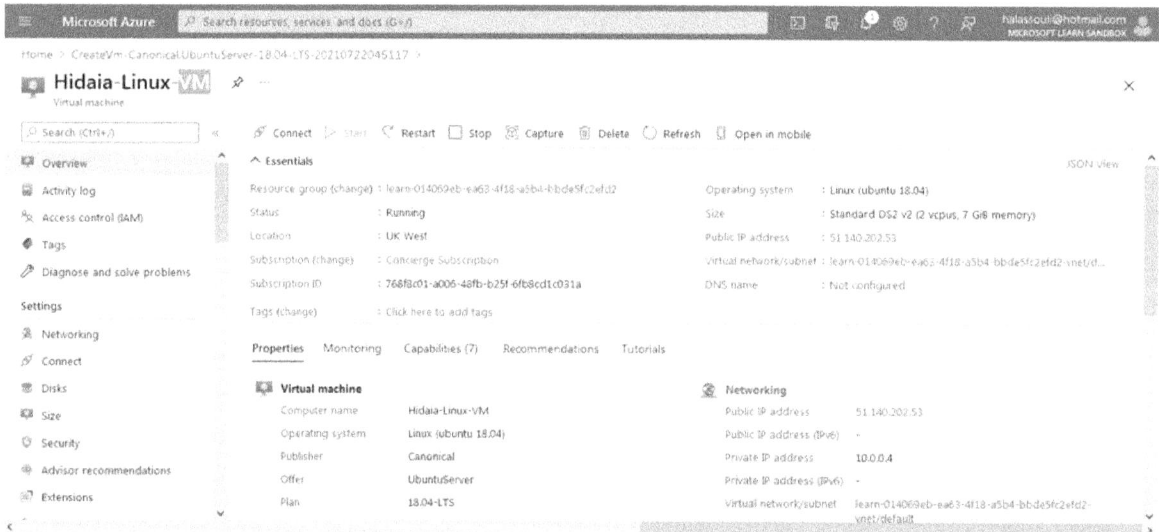

4. In the networking section, you will see the list of firewall rules for the inbound connection ports that are allowed in the server

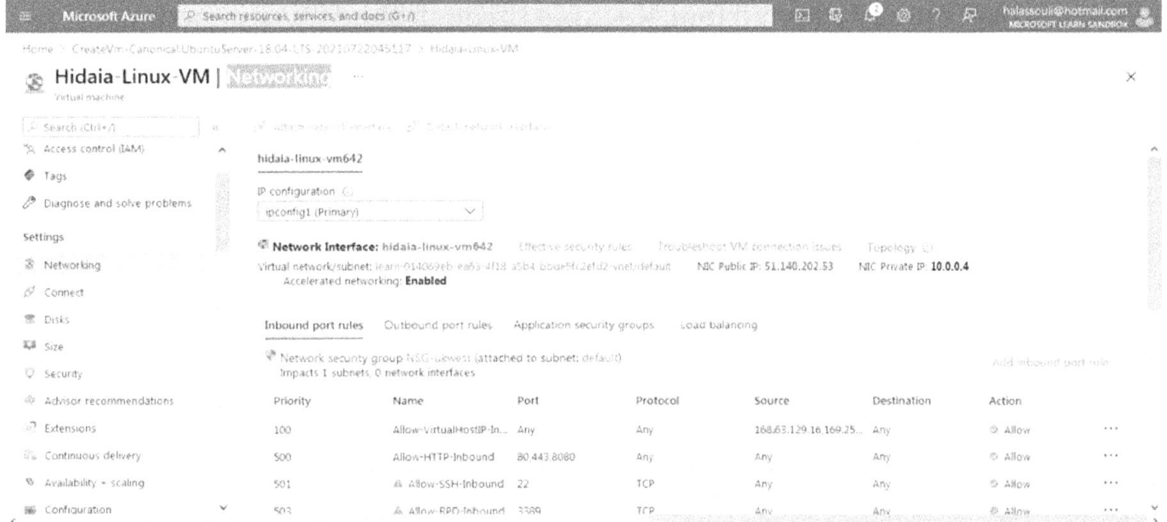

- You can add new rule to allow VNC inbound connection to the Linux Machine VNC service port 5901 to allow connection to the VNC service in the VM machine.

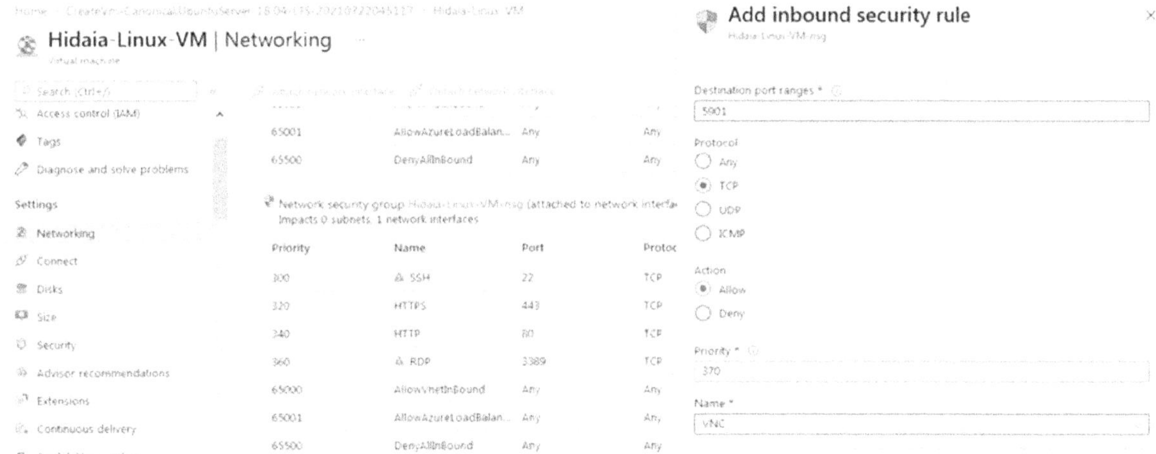

5. Restart the Linux virtual machine.

c) Connection to the Linux Virtual Machine through the SSH connection

1. From the Virtual machine dashboard, restart the machine and choose connect to the Linux machine. You will be given three option of connections based on the machine whether Linux or Windows machine. For windows machine, you connect through RDP connection. And for Linux machine you connect through SSH connection. There is third option which is BASTION connection which will not be covered here.

2. For SSH connection to Linux virtual machine, you will be given some guide to connect to server through SSH connection.

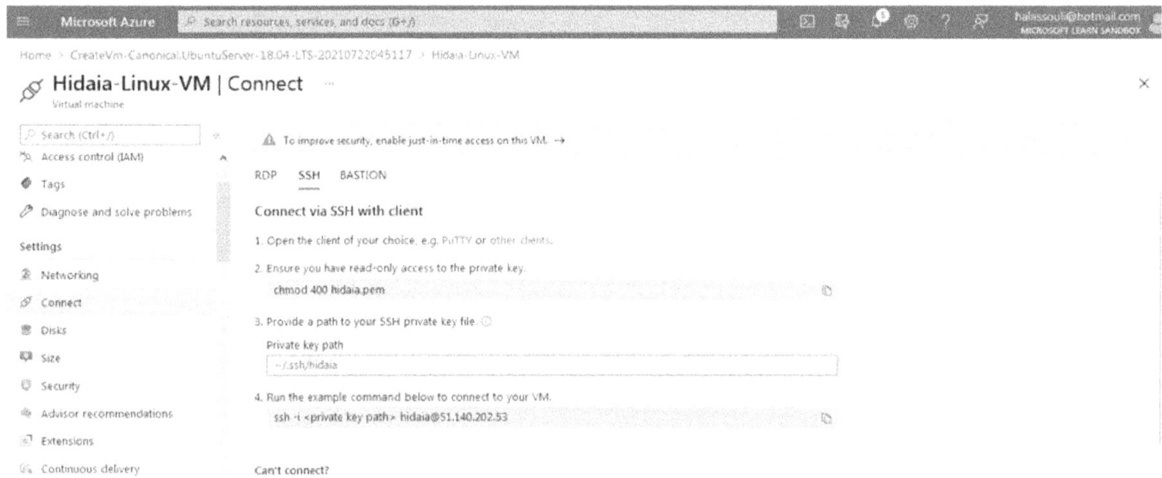

3. I connected to the Linux virtual machine through SSH connection using PuTTY tool using the username and password I setup when I launched the VM. This is the simplest way to connect to server

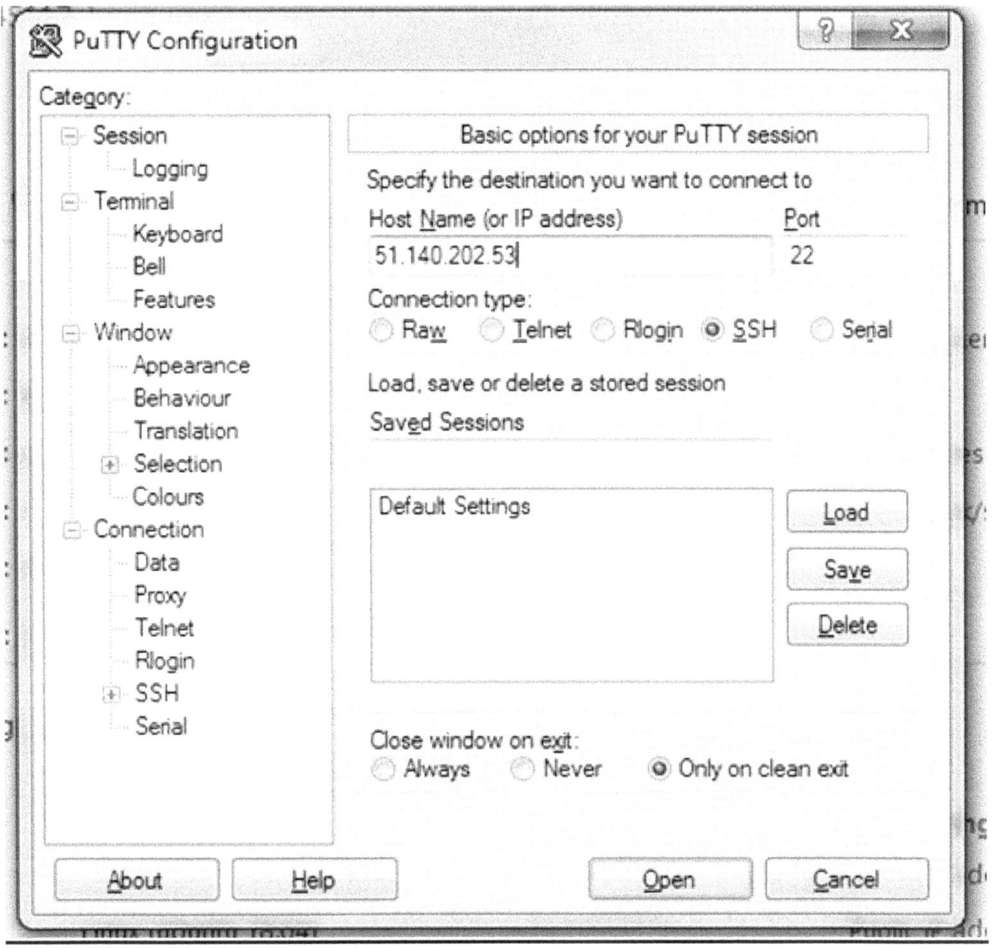

4. Fellow the steps in the previous sections to install XFCE as graphical user interface (GUI) and connect to the Linux Machine instance through RDP and VNC connection:

5. Install XRDP server

```
$ sudo apt-get  update

$ sudo apt-get  upgrade

$ sudo apt-get install -y xrdp
```

6. Start XRDP server

```
$ sudo service xrdp restart

$ sudo ufw allow 3389/tcp
```

7. Install XFCE packages

$ sudo apt-get install xfce4 xfce4-goodies

$ sudo apt-get install xfce4-terminal

8. Install firefox internet browser

$ sudo apt install firefox

9. If you prefer Google chrome browser, you can download chromium-browser through the following command:
$ sudo apt install chromium-browser

10. When you connect to the Linux server through RDP connection, better to add a user and reset the root password so you connect to server by either root account or user account.

- Create username, as example hasooly using add user command. Specify the password and user information.

$ sudo adduser hasooly

```
student-00-a3ac361471ed@instance-1:~$ sudo adduser hasooly
Adding user `hasooly' ...
Adding new group `hasooly' (1003) ...
Adding new user `hasooly' (1002) with group `hasooly' ...
Creating home directory `/home/hasooly' ...
Copying files from `/etc/skel' ...
Enter new UNIX password:
Retype new UNIX password:
passwd: password updated successfully
Changing the user information for hasooly
Enter the new value, or press ENTER for the default
        Full Name []: hidaia
        Room Number []:
        Work Phone []:
        Home Phone []:
        Other []:
Is the information correct? [Y/n] y
student-00-a3ac361471ed@instance-1:~$
```

- To update also the root user, connect to the server using super root account. Then change the password of the root user by passwd command

$ sudo su

passwd

Give the new password of the root

```
root@instance-1:/home/student-00-a3ac361471ed# passwd
Enter new UNIX password:
Retype new UNIX password:
passwd: password updated successfully
root@instance-1:/home/student-00-a3ac361471ed#
```

11. Using windows remote access client, connect to the server by giving the IP address and the username and password in the server. In my case

IP: (Provide public IP of the instance)

Username: hasooly

Password:

You will be logged on to the machine.

12. In case you need a VNC server to interact with desktop environment. You can use vnc4server, you can install your favorite one:

$ sudo apt-get install vnc4server

Or you can tightvncserver

$ sudo apt install -y tightvncserver

I used here tightvncserver

$ sudo apt install -y tightvncserver
$ sudo ufw allow 5901:5910/tcp

13. Start the vnc server, you'll then be prompted to create and verify a new password:

$ tightvncserver

```
student-00-621a4ce59b11@instance-1:~$ vncserver

You will require a password to access your desktops.

Password:
Verify:
xauth:  file /home/student-00-621a4ce59b11/.Xauthority does not exist

New 'X' desktop is instance-1:1

Creating default startup script /home/student-00-621a4ce59b11/.vnc/xstartup
Starting applications specified in /home/student-00-621a4ce59b11/.vnc/xstartup
Log file is /home/student-00-621a4ce59b11/.vnc/instance-1:1.log
```

14. If everything went fine your VNC server is now running and listening on port 5901. You can verify this with netcat from the Google Compute Engine instance:

$ nc localhost 5901

RFB 003.008

15. Make sure to create firewall rule to allow incoming Tcp connection to the VNC server port 5901 following the steps in previous sections.

16. We now need to kill the session we just created and make a tweak to the startup script for VNC Server to make it work properly. If we don't perform this step then all we will see is a grey cross-hatched screen with an "X" cursor and/or a grey screen with a Terminal Session, depending on the Ubuntu version. So, type the following command to kill the session:

$ vncserver -kill :1

17. A startup script is automatically generated by the shell, as shown in the following screenshot. This startup script can be accessed and edited by copying and pasting its `PATH` in the following manner: In my case I got the following xtratup script and log file.

Starting applications specified in /home/student-00-621a4ce59b11/.vnc/xstartup
Log file is /home/student-00-621a4ce59b11/.vnc/instance-1:1.log

Now open the file we need to edit:

$ vim .vnc/xstartup

The xstartup file will look when opened first time

```
#!/bin/sh

xrdb $HOME/.Xresources
xsetroot -solid grey
#x-terminal-emulator -geometry 80x24+10+10 -ls -title "$VNCDESKTOP Desktop" &
#x-window-manager &
# Fix to make GNOME work
export XKL_XMODMAP_DISABLE=1
/etc/X11/Xsession
```

18. Press the [Insert] key ("i" in Ubuntu) once (this will switch us into "edit" mode) and then edit the script so it ends up looking like this:

#!/bin/bash

xrdb $HOME/.Xresources

startxfce4 &

When you're done editing the. vnc/xstartup file for your particular version of Ubuntu press the [Esc] key once and type the following to save the changes and bring you back to the command line:

```
:wq
```

19. You must kill the vncservor, then start the server again

```
$ vncserver -kill :1
$  vncserver
```

20. Install VNC client in your computer. There are many options available, one of them is RealVNC Viewer. Install one but don't try to connect to your server just yet: it will fail as the firewall rules don't allow it. The download link for RealVNC Viewer

https://www.realvnc.com/en/connect/download/viewer/

21. Open your VNC viewer and connect to the IP of your Compute Engine instance on port 5901.

22. Your Desktop environment is working. You can use Firefox browser for internet.

VNC Viewer
File View Help
Enter a VNC Server address or search
Sign in...
34.89.94.3:5901
34.89.94.3:5901 - VNC Viewer
Connecting to 34.89.94.3:5901...
Stop

d) Other connection methods to the Linux Virtual Machine using the SSH key:

1. Download the SSH key of the Linux VM. Create an SSH connection with the VM. If you are on a Mac or Linux machine, open a Bash prompt. If you are on a Windows machine, open a PowerShell prompt. At your prompt, open an SSH connection to your virtual machine. Replace the IP address with the one from your VM, and replace the path to the. pem with the path to where the key file was downloaded.

> ssh -i .\Downloads\SSH-key.pem azureuser@(IP address)

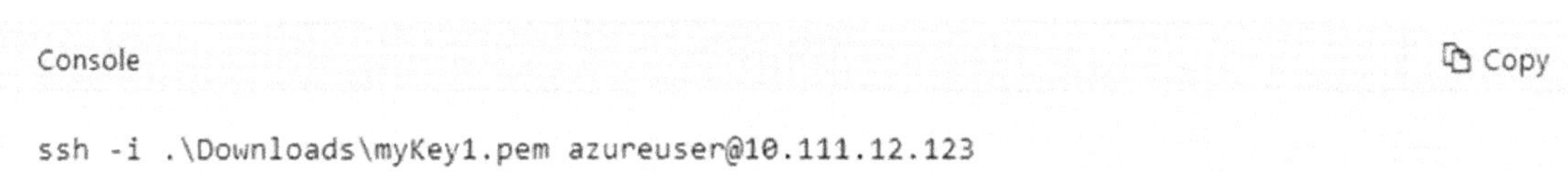

2. Install web server

- To see your VM in action, install the NGINX web server. From your SSH session, update your package sources and then install the latest NGINX package.

  >sudo apt-get -y update

  >sudo apt-get -y install nginx

Bash Copy

```
sudo apt-get -y update
sudo apt-get -y install nginx
```

- Use a web browser of your choice to view the default NGINX welcome page. Type the public IP address of the VM as the web address. The public IP address can be found on the VM overview page or as part of the SSH connection string you used earlier.

10.111.12.123

Welcome to nginx!

If you see this page, the nginx web server is successfully installed and working. Further configuration is required.

For online documentation and support please refer to nginx.org.
Commercial support is available at nginx.com.

Thank you for using nginx.

10. Quick guide to create a Linux Virtual Machine in Amazon AWS:

More than 85 products on AWS using the free tier. Three different types of free offers are available depending on the product used. See the details of each product from https://aws.amazon.com/free/

Note that: I don't have balance in my Amazon AWS account, so I just explain the steps without testing the server as I could not launch the server without having credits in my Amazon AWS account.

1. Register account at https://aws.amazon.com/

2. Go to AWS console https;//console.aws.amazon.com. When you log in you will get this screen

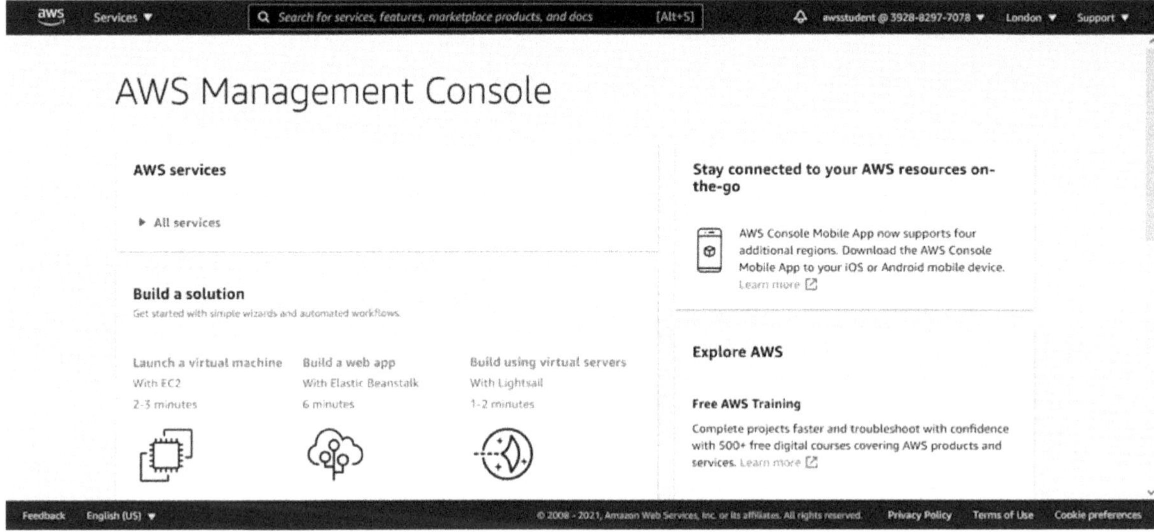

3. Choose AWS services and you will get all AWS services. Here the list of services:

AWS services

Find Services
You can enter names, keywords or acronyms.

Example: Relational Database Service, database, RDS

▼ All services

Compute
EC2
Lightsail
Lambda
Batch
Elastic Beanstalk
Serverless Application Repository
AWS Outposts
EC2 Image Builder

Containers
ECR
Elastic Container Service
Elastic Kubernetes Service

Storage
S3
EFS
FSx
S3 Glacier
Storage Gateway
AWS Backup

Database
RDS
DynamoDB
ElastiCache
Neptune
Amazon QLDB
Amazon DocumentDB
Amazon Keyspaces
Amazon Timestream

Migration & Transfer
AWS Migration Hub
Application Discovery Service
Database Migration Service
Server Migration Service
AWS Transfer Family
AWS Snow Family
DataSync

Networking & Content Delivery
VPC
CloudFront
Route 53
API Gateway
Direct Connect
AWS App Mesh
AWS Cloud Map
Global Accelerator

Developer Tools
CodeStar
CodeCommit
CodeArtifact
CodeBuild
CodeDeploy
CodePipeline
Cloud9
X-Ray

Customer Enablement
AWS IQ
Support
Managed Services
Activate for Startups

Robotics
AWS RoboMaker

Blockchain
Amazon Managed Blockchain

Satellite
Ground Station

Quantum Technologies
Amazon Braket

Management & Governance
AWS Organizations
CloudWatch
AWS Auto Scaling
CloudFormation
CloudTrail
Config
OpsWorks
Service Catalog
Systems Manager
AWS AppConfig
Trusted Advisor
Control Tower
AWS License Manager
AWS Well-Architected Tool
Personal Health Dashboard
AWS Chatbot
Launch Wizard
AWS Compute Optimizer
Resource Groups & Tag Editor

Media Services
Kinesis Video Streams
MediaConnect
MediaConvert
MediaLive
MediaPackage
MediaStore
MediaTailor
Elemental Appliances & Software
Amazon Interactive Video Service
Elastic Transcoder

Machine Learning
Amazon SageMaker
Amazon Augmented AI
Amazon CodeGuru
Amazon Comprehend
Amazon Forecast
Amazon Fraud Detector
Amazon Kendra
Amazon Lex
Amazon Personalize
Amazon Polly
Amazon Rekognition
Amazon Textract
Amazon Transcribe
Amazon Translate
AWS DeepComposer
AWS DeepLens
AWS DeepRacer

Analytics
Athena
Amazon Redshift
EMR
CloudSearch
Elasticsearch Service
Kinesis
QuickSight
Data Pipeline
AWS Data Exchange
AWS Glue
AWS Lake Formation
MSK

Security, Identity, & Compliance
IAM
Resource Access Manager
Cognito
Secrets Manager
GuardDuty
Inspector
Amazon Macie
AWS Single Sign-On
Certificate Manager
Key Management Service
CloudHSM
Directory Service
WAF & Shield
AWS Firewall Manager
Artifact
Security Hub
Detective

AWS Cost Management
AWS Cost Explorer
AWS Budgets
AWS Marketplace Subscriptions

Front-end Web & Mobile
AWS Amplify
Mobile Hub
AWS AppSync
Device Farm

AR & VR
Amazon Sumerian

Application Integration
Step Functions
Amazon AppFlow
Amazon EventBridge
Amazon MQ
Simple Notification Service
Simple Queue Service
SWF

Customer Engagement
Amazon Connect
Pinpoint
Simple Email Service

Business Applications
Alexa for Business
Amazon Chime
WorkMail
Amazon Honeycode

End User Computing
WorkSpaces
AppStream 2.0
WorkDocs
WorkLink

Internet of Things
IoT Core
FreeRTOS
IoT 1-Click
IoT Analytics
IoT Device Defender
IoT Device Management
IoT Events
IoT Greengrass
IoT SiteWise
IoT Things Graph

Game Development
Amazon GameLift

4. Choose EC2 Service option, https://console.aws.amazon.com/ec2

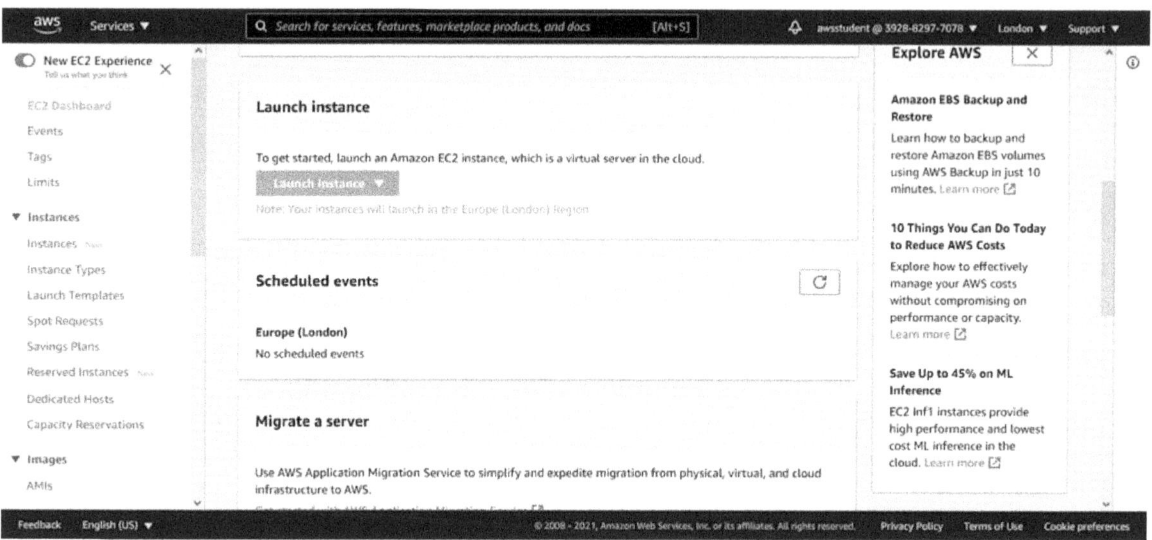

5. Then choose launch instance. Choose your suitable Amazon Linux Machine Image. In my example I chose Ubuntu Server 18.04 LTS (HVM

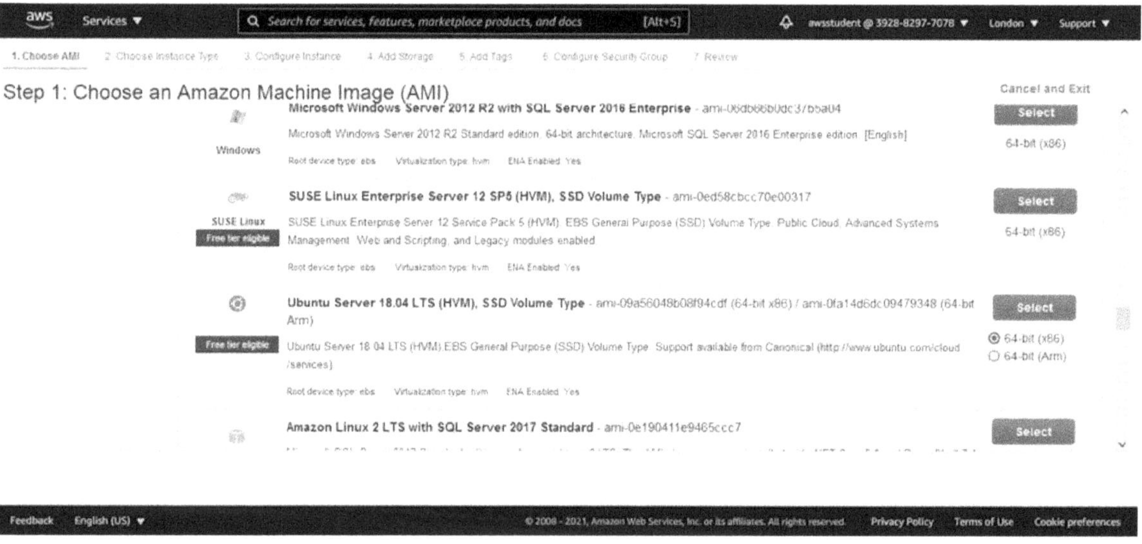

6. In my example I chose Ubuntu Server 18.04 LTS (HVM). Choose an Instance Type. Amazon EC2 provides a wide selection of instance types optimized to fit different use cases. Instances are virtual servers that can run applications. They have varying combinations of CPU, memory, storage, and networking capacity, and give you the flexibility to choose the appropriate mix of resources for your applications. When you selected the machine, you will see one of the instances is highlighted as 12 micro free tires eligible.

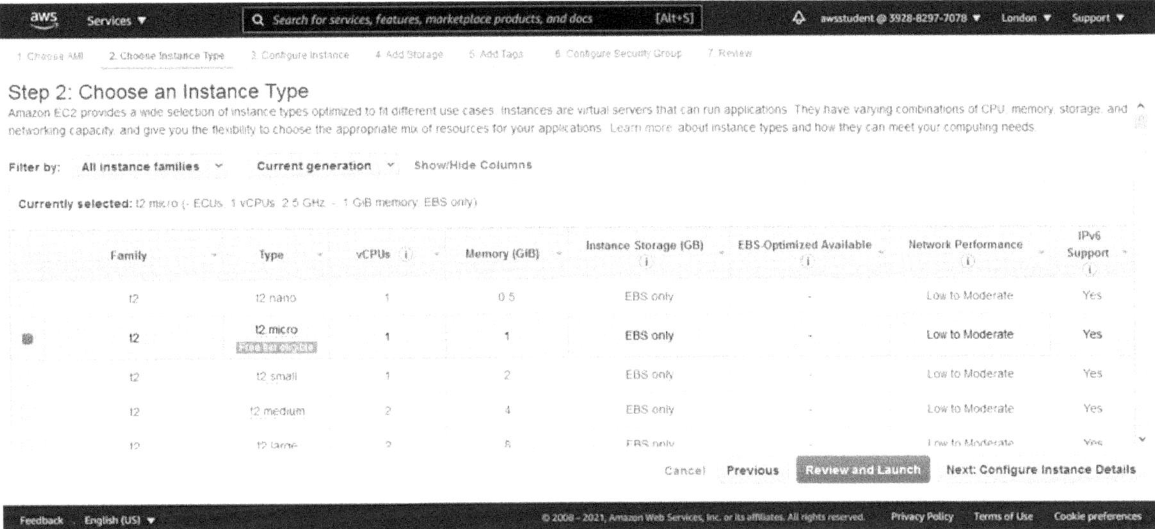

7. Next page, you must configure Instance Details. Configure the instance to suit your requirements. You can launch multiple instances from the same AMI, request Spot instances to take advantage of the lower pricing, assign an access management role to the instance, and more. In my example I left the default configuration for the machine.

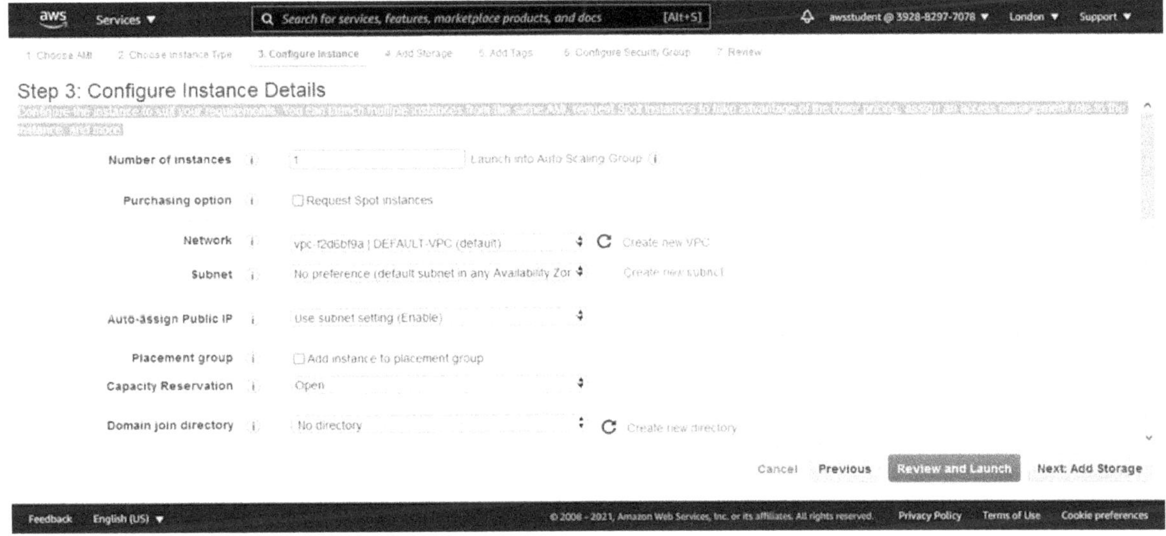

8.In the next page you must Add Storage.

Your instance will be launched with the following storage device settings. You can attach additional EBS volumes and instance store volumes to your instance, or edit the settings of the root volume. You can also attach additional EBS volumes after launching an instance, but not instance store volumes. In my example, I chose 30 GiB storage for the machine.

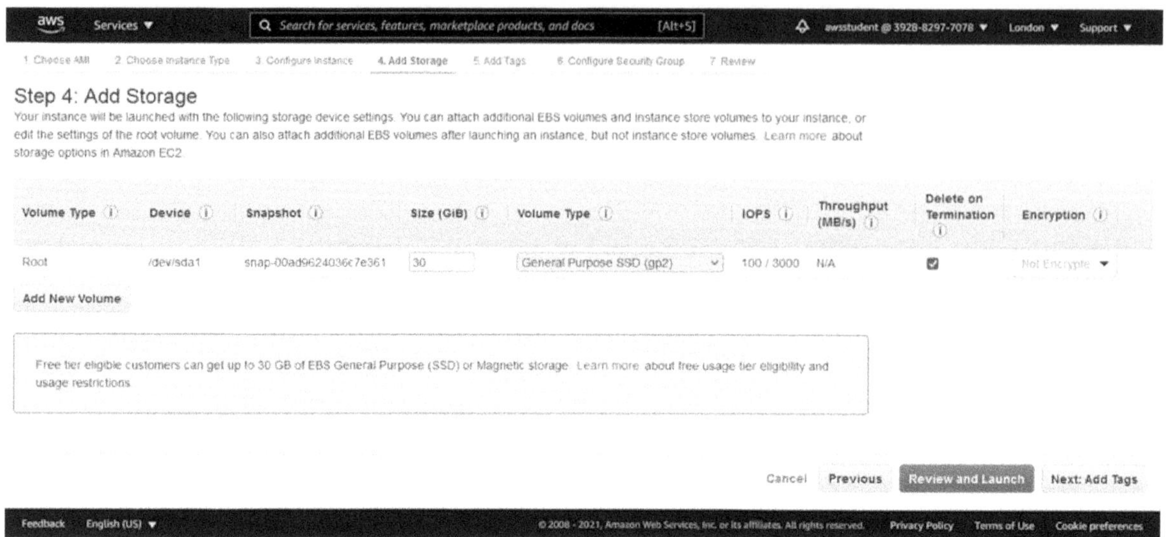

9. Click next to add tags. A tag consists of a case-sensitive key-value pair. For example, you could define a tag with key = Name and value = Webserver. A copy of a tag can be applied to volumes, instances or both. Tags will be applied to all instances and volumes

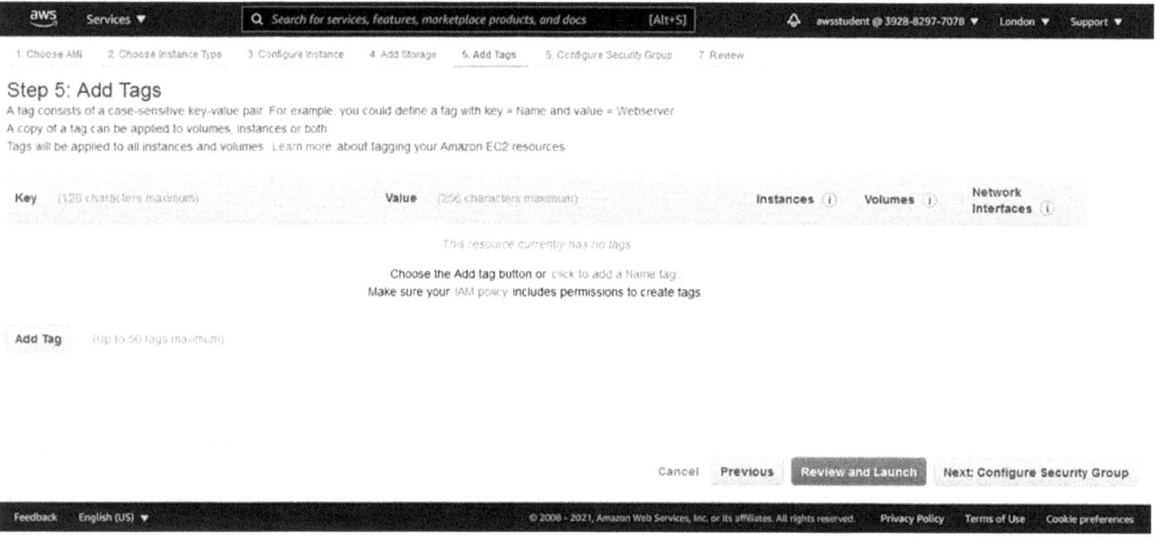

10. Click next, you will be forwarded to a page that configures Configure Security Group. A security group is a set of firewall rules that control the traffic for your instance. On this page, you can add rules to allow specific traffic to reach your instance. For example, if you want to set up a web server and allow Internet traffic to reach your instance, add rules that allow unrestricted access to the HTTP and HTTPS ports. You can create a new security group or select from an existing one below. The default rule in Linux machine is to allow incoming connections to SSH TCP port 22 of the machine. In my example I created rules that allows incoming connections to the server to RDP TCP port 2289 and VNC TCP port 5901

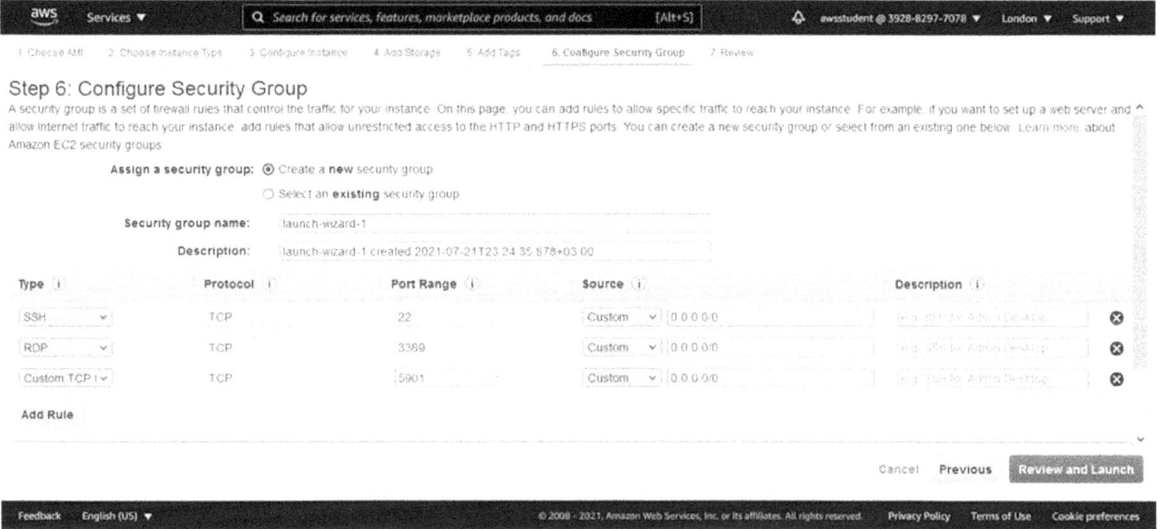

11. In last step, you review and launch. Please review your instance launch details. You can go back to edit changes for each section.

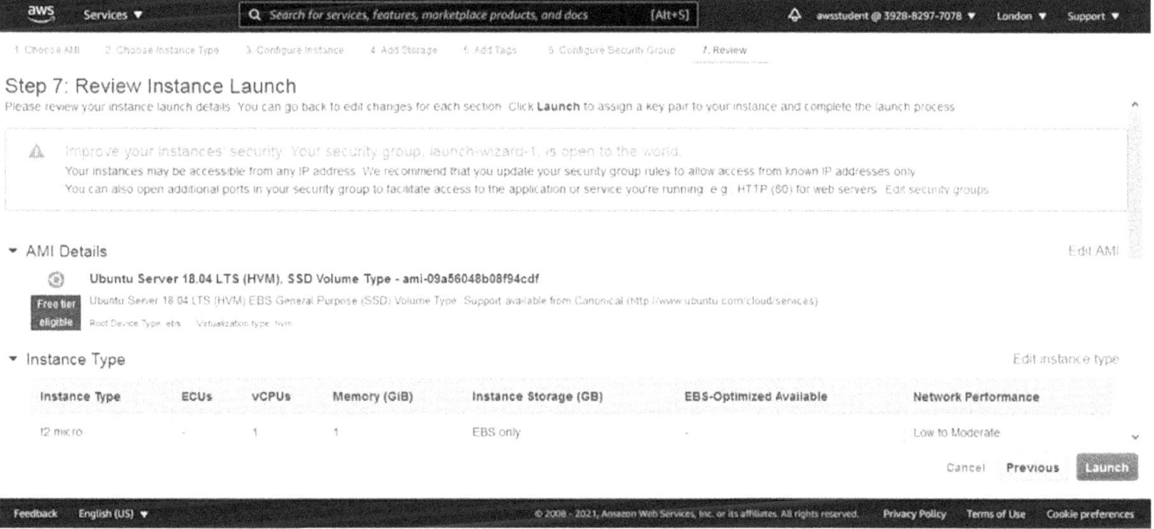

12. Click Launch to assign a key pair to your instance and complete the launch process. A key pair consists of a public key that AWS stores, and a private key file that you store. Together, they allow you to connect to your instance securely. For Windows AMIs, the private key file is required to obtain the password used to log into your instance. For Linux AMIs, the private key file allows you to securely SSH into your instance. In my case I created new key file and downloaded it to use it to login to the server.

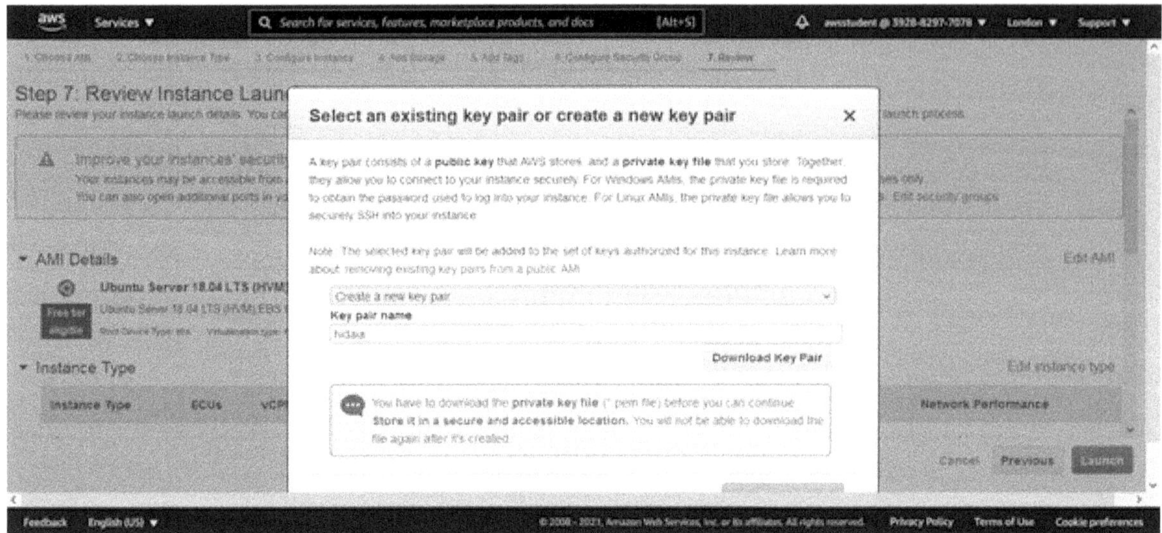

13. Then the instance will be created and running

14. Then choose action tab, and choose connect.

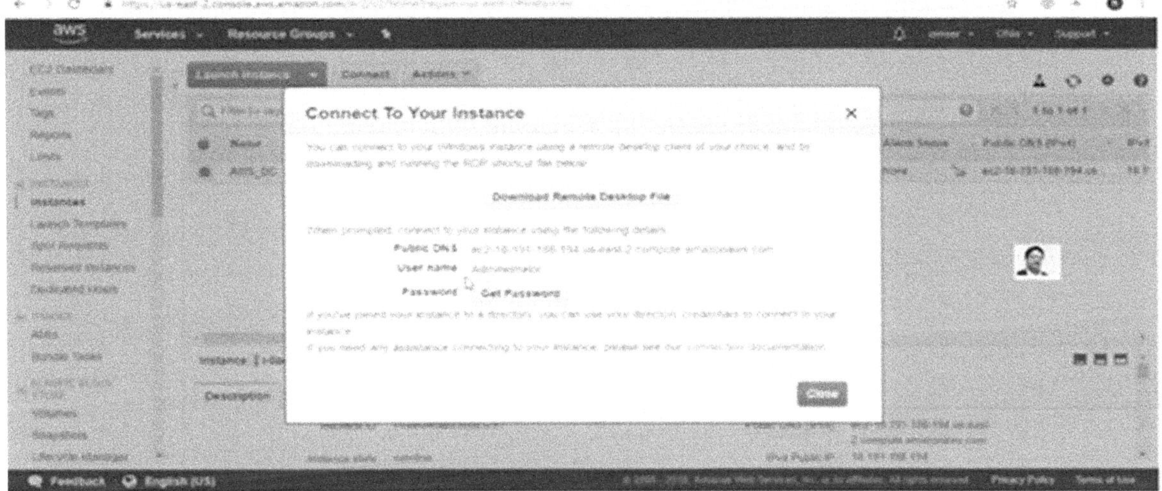

15. Then you can connect through PuTTY tool to the server through SSH connection by giving the IP address and the SSH key file for login.

16. Note that: I don't have balance in my Amazon AWS account, so I just explain the steps without testing the server as I could not launch the server without having credits in my Amazon AWS account.

www.ingramcontent.com/pod-product-compliance
Ingram Content Group UK Ltd.
Pitfield, Milton Keynes, MK11 3LW, UK
UKHW061702190726
13853UKWH00008B/2354

9 798211 945999